普通高等院校计算机类专业规划教材

Java 程序设计

（中英双语版）

主　编　田玉昆　陈　伟　谢文兰
副主编　徐光明　崔彦君　魏文芬
参　编　黄国范　张艳梅　曾　兰

JAVA PROGRAMMING

中国铁道出版社有限公司
CHINA RAILWAY PUBLISHING HOUSE CO., LTD.

内 容 简 介

随着互联网技术的发展及应用的普及，众多高等院校的计算机和非计算机专业均将 Java 程序设计课程作为程序设计的入门课程或者程序设计的进阶课程。

全书共 11 章，内容主要为：Introduction（简介）；Java Foundation（Java 基础）；Classes and Objects（类和对象）；Packages（包）；Inheritance（继承）；Abstract Class，Interface（抽象类、接口）；Generics and Collections（泛型与集合）；Exception Handling（异常处理）；I/O（输入/输出）；Multi-threading（多线程）；Networking（网络）。

本书适合作为计算机及计算机相近专业的 Java 程序设计课程的教材，也可作为普通高等院校面向对象程序设计语言课程的教材和参考书，还可作为软件开发人员及其他有关人员自学的参考教材或培训教材。

图书在版编目（CIP）数据

Java 程序设计：中英双语版 / 田玉昆，陈伟，谢文兰 主编 .—北京：中国铁道出版社有限公司，2019.9（2022.1重印）
普通高等院校计算机类专业规划教材
ISBN 978-7-113-26185-6

Ⅰ . ① J… Ⅱ . ①田… ②陈… ③谢… Ⅲ . ① JAVA 语言－程序设计－高等学校－教材－汉、英 Ⅳ . ① TP312.8

中国版本图书馆 CIP 数据核字（2019）第 181377 号

书　　名	Java 程序设计（中英双语版）
作　　者	田玉昆　陈　伟　谢文兰

策　　划	唐　旭	编辑部电话	（010）63549508
责任编辑	唐　旭　李学敏		
封面设计	尚明龙		
封面制作	刘　颖		
责任校对	张玉华		
责任印制	樊启鹏		

出版发行：中国铁道出版社有限公司（100054，北京市西城区右安门西街 8 号）
网　　址：http://www.tdpress.com/51eds/
印　　刷：三河市航远印刷有限公司
版　　次：2019 年 9 月第 1 版　2022 年 1 月第 2 次印刷
开　　本：787 mm×1 092 mm　1/16　印张：14.5　字数：346 千
书　　号：ISBN 978-7-113-26185-6
定　　价：45.00 元

版权所有　侵权必究

凡购买铁道版图书，如有印制质量问题，请与本社教材图书营销部联系调换。电话：（010）63550836
打击盗版举报电话：（010）63549461

前言 PREFACE

  Java 语言具有面向对象、跨平台、安全性、多线程等特点，这使得 Java 成为许多应用系统的理想开发语言。Java 应用程序运行在众多计算机、手机和智能卡上，并为无数的可兼容的应用服务器提供了功能强大的平台。未来发展中，Java 将成为 IT 从业者必须掌握的一门语言，而且在金融、电信、制造、银行、移动通信、电力、交通、市政服务、政务管理等行业的应用日益广泛。为了满足高校对 Java 程序设计语言课程和双语课程教材的需求，满足普通高等院校学生和编程爱好者希望快速掌握 Java 程序开发的基本知识和开发原理的需求，我们编写了本书。

  本书采用中英文对照的方式，将较难理解的英文叙述翻译成对应的汉语（部分简单的英文内容未作翻译），为读者在基本概念的理解上提供方便。对于书中复杂的知识点则辅以简单、清晰的代码，并将运行结果附于代码之后，方便读者比较学习。

  全书共 11 章：Chapter 1 Introduction（简介），主要介绍 Java 语言的特点、工作环境与 Java 的基本工作原理，使读者对 Java 语言有一个基本的了解；Chapter 2 Java Foundation（Java 基础），主要介绍数据类型、常量与变量、表达式和流程控制语句等；Chapter 3 Classes and Objects（类和对象），主要介绍 Java 类与对象的基本概念、构造方法、对象的初始化等内容，包括 Java 面向对象的特征等；Chapter 4 Packages（包），介绍了包的概念和组织结构，在此基础上，讲述了包的创建、导入及使用；Chapter 5 Inheritance（继承），继承是 OOP 的一个重要内容，本章在类的基础上介绍了继承的基本概念及构造方法的调用，对不同类型访问控制符的作用进行了比较，最后对 OOP 的另一个重要内容多态作了简要介绍；Chapter 6 Abstract Class, Interface（抽象类、接口），阐述了抽象方法、抽象类的定义与使用，接口的定义与实现，接口与抽象类的比较以及内部类的定义与使用等内容；Chapter 7 Generics and Collections（泛型与集合），介绍了泛型的概念、泛型类、泛型接口与泛型方法的定义与使用，以及集合类的概念、框架结构和几种常用的容器的使用方法；Chapter 8 Exception Handling（异常处理），介绍了异常产生的原因、异常的类型，以及如何抛出和处理异常；Chapter 9 I/O（输入/输出），介绍了输入/输出流的概念，以及如何通过输入流读取数据，向输出流写入数据；Chapter 10 Multithreading（多线程），阐述了多线程的概念，线程与进程的区别，线程的生命周期与创建方法、线程同步与线程间的通信等；Chapter 11 Networking（网络），介绍了 Java 网络编程技术，描述了 Java 的网络编程功能。全书对不易理解的内容的细致分析，从学生学习的角度阐述核心概念，具有很强的实用性和适用性。

本书由田玉昆、陈伟、谢文兰任主编，徐光明、崔彦君、魏文芬任副主编，黄国范、张艳梅、曾兰参与编写。吴学曼对本书的编写给予了专业性的指导，在此表示感谢。

本书的出版得到了广东培正学院教材立项支持，在此表示衷心的感谢。

由于编者水平有限、时间仓促，书中疏漏或不足之处在所难免，请广大读者批评指正。

编　者
2019 年 6 月

目 录
CONTENTS

Chapter 1　Introduction（简介） 1
　1.1　Characteristics of Java（Java 的特点） 1
　1.2　Developing tools（开发工具） 2
　　1.2.1　Interpreted language vs compiled language（解释性语言与编译性语言）.......... 2
　　1.2.2　Java Virtual Machine（虚拟机）... 3
　　1.2.3　Java Runtime Environment（运行环境）.................................. 4
　　1.2.4　Java Development Kit（开发工具）....................................... 4
　1.3　A simple Java program（一个简单的 Java 程序）................................ 5
　　1.3.1　How to run a Java program?（如何运行 Java 程序？）..................... 5
　　1.3.2　Rules of naming a Java source file（Java 源文件的命名规则）............. 6
　　1.3.3　Rules of naming classes, variables and methods（类、变量和方法的命名规则）... 7
　Exercises 8

Chapter 2　Java Foundation（Java 基础）... 9
　2.1　Primitive data types（基本数据类型）.. 9
　　2.1.1　Integer and floating point types（整型与浮点型）....................... 10
　　2.1.2　Boolean type（布尔型）.. 11
　　2.1.3　Character type（字符型）.. 11
　2.2　Reference types（引用型）... 12
　　2.2.1　A class is a data type（类是数据类型）................................. 13
　　2.2.2　A class type variable is a reference（类类型的变量是引用）............. 14
　　2.2.3　Interface type（接口类型）.. 16
　2.3　Identifiers（标识符）... 16
　2.4　Default values of fields（成员变量的默认值）................................. 16
　2.5　Where data store?（数据存储在何处？）....................................... 18
　2.6　Operators（运算符）... 19
　　2.6.1　Arithmetic operators（算术运算符）..................................... 19
　　2.6.2　Logical operators（逻辑运算符）.. 19

2.6.3　Bitwise operators（位运算符）..20

2.6.4　Left shift (<<) and signed right shift (>>) operators（移位运算符）..............21

2.6.5　Assignment and conditional operators（赋值运算符和条件运算符）.............22

2.6.6　String operator "+" and "+="（字符串运算符"+"和"+="）..............22

2.6.7　Special operators（特殊运算符）..23

2.7　Casting（类型转换）..23

2.7.1　Widening and narrowing（拓宽与缩窄）...23

2.7.2　Char, byte and short produce int results（Char, byte 和 short 转换为 int 型）.........25

2.8　Flowing control（流程控制）..25

2.8.1　Basic controlling statements（基本控制语句）...26

2.8.2　Foreach statement（foreach 语句）..26

2.9　Arrays（数组）..28

2.9.1　Define arrays（定义数组）...28

2.9.2　Initialize arrays（初始化数组）..28

2.9.3　Arrays act as arguments of methods（数组做方法的参数）.........................29

2.10　Command line arguments（命令行参数）...30

Exercises..31

Chapter 3　Classes and Objects（类和对象）..33

3.1　Concepts of OOP（面向对象的概念）..33

3.1.1　Everything is an Object（万物皆对象）...33

3.1.2　Defining classes（定义类）..34

3.2　Useful classes（常用类）...34

3.3　Method overloading（方法重载）..39

3.4　Constructors（构造方法）..40

3.5　Default constructor（默认构造方法）..43

3.6　Static fields and methods（静态成员变量与静态方法）..45

3.7　This keyword（this 关键字）..51

3.7.1　A non-static method has a hidden "this"（隐藏参数 this 的非静态方法）.........51

3.7.2　A static method has no argument "this"（没有 this 的静态方法）..............54

3.7.3　Calling constructors form constructors（在构造方法里调用其他构造方法）.......56

3.8　Variable argument lists（可变参数列表）..56

3.9　Garbage collection（垃圾回收）..58

3.10　Enum type（枚举类型）...60

Exercises..61

Chapter 4 Packages（包） ... 64
4.1 Concept of packages（包的概念） ... 64
4.2 Java library and its package structure（类库与 Java 类的包组织结构） ... 65
4.3 Create packages（创建包） ... 66
4.4 Import packages（导入包） ... 68
4.5 Package java.lang（java.lang 包） ... 70
4.6 Useful classes in package java.lang（java.lang 包中常用的类） ... 71
4.6.1 Object class and its toString method（Object 类和它的 toString 方法） ... 71
4.6.2 System class（系统类） ... 72
Exercises ... 72

Chapter 5 Inheritance（继承） ... 74
5.1 What is inheritance?（什么是继承？） ... 74
5.1.1 Root class Object（根类对象） ... 75
5.1.2 Defining subclasses（定义子类） ... 76
5.2 Super keyword（super 关键字） ... 78
5.2.1 Super corresponding to default constructor（默认构造方法的 super） ... 78
5.2.2 Super corresponding to constructors with arguments（有参构造方法的 super） ... 80
5.3 Order of constructor calls（构造方法的调用次序） ... 82
5.4 Final keyword（final 关键字） ... 84
5.4.1 Final fields（final 成员） ... 84
5.4.2 Final arguments（常参数） ... 86
5.4.3 Final methods（常方法） ... 86
5.5 Access specifiers（访问说明符） ... 86
5.6 Polymorphism（多态） ... 91
5.6.1 Method overriding（方法重写） ... 92
5.6.2 Upcasting and dynamic polymorphism（升级转换与动态多态） ... 93
5.6.3 Referring to a member of the super class by super keyword（用 super 指向基类成员） ... 96
5.6.4 Hiding fields and static methods of the base class（隐藏静态方法和 fields 的基类） ... 98
Exercises ... 100

Chapter 6 Abstract Class and Interface（抽象类和接口） ... 104
6.1 Abstract class（抽象类） ... 105
6.1.1 Abstract method（抽象方法） ... 105

6.1.2 Abstract class（抽象类） 105
6.2 Interface（接口） 106
　6.2.1 Introduction（简介） 106
　6.2.2 Defining interfaces（接口的定义） 107
　6.2.3 Implementation of interfaces（接口的实现） 108
　6.2.4 Comparation of interfaces and abstract classes（接口与抽象类的比较） 109
6.3 Inner class（内部类） 110
　6.3.1 Members in inner class（内部类成员） 110
　6.3.2 Local inner class（局部内部类） 112
　6.3.3 Anonymous inner class（匿名内部类） 113
Exercises 115

Chapter 7　Generics and Collections（泛型与集合） 117

7.1 Generics（泛型） 117
　7.1.1 Concept of generics（泛型的概念） 117
　7.1.2 Generic classes（泛型类） 118
　7.1.3 Type parameters use "extends" and "super" keywords（类型参数中使用 extends 和 super 关键字） 121
　7.1.4 Wildcard in type parameters（类型参数里的通配符） 122
7.2 Generic interfaces（泛型接口） 122
7.3 Generic methods（泛型方法） 122
7.4 Collection classes（集合类） 123
　7.4.1 Concept of collection (container)classes［集合（容器）类的概念］ 124
　7.4.2 The hierarchy of the collection framework（集合框架的层次结构） 124
7.5 List（列表） 126
7.6 Queue（队列） 131
7.7 Set（集合） 131
7.8 Map（映射） 133
Exercises 135

Chapter 8　Exception Handing（异常处理） 137

8.1 Concepts of exception（异常的概念） 137
　8.1.1 What is an exception?（什么是异常？） 137
　8.1.2 How to deal with exceptions?（如何处理异常？） 138
8.2 Exception classes（异常类） 140

8.2.1　Error class（Error 类） 141
  8.2.2　Exception class（Exception 类） 141
 8.3　Catch and deal with an exception（捕获与处理异常） 144
 8.4　Throw an exception（抛出一个异常） 148
 8.5　Define your own exceptions（用户自定义异常类） 152
 Exercises 153

Chapter 9　I/O（输入/输出） 155

 9.1　Concept of I/O stream（输入/输出流的概念） 155
 9.2　Byte stream（字节流） 156
  9.2.1　InputStream/OutputStream（InputStream/OutputStream 类） 157
  9.2.2　FileInputStream/FileOutputStream（FileInputStream/FileOutputStream 类） 159
  9.2.3　FilterInputStream/FilterOutputStream（FilterInputStream/
       FilterOutputStream 类） 163
  9.2.4　DataInputStream/DataOutputStream（DataInputStream/
       DataOutputStream 类） 164
  9.2.5　BufferedInputStream/BufferedOutputStream（BufferedInputStream/
       BufferedOutputStream 类） 165
  9.2.6　PrintStream（PrintStream 类） 166
 9.3　Character streams（字符流） 169
  9.3.1　Text file vs Binary file（文本文件与二进制文件） 170
  9.3.2　Reader/Writer（Reader/Writer 类） 171
  9.3.3　InputStreamReader/OutputStreamWriter（InputStreamReader/
       OutputStreamWriter 类） 172
  9.3.4　FileReader/FileWriter（FileReader/FileWriter 类） 175
  9.3.5　BufferReader/BufferWriter（BufferReader/BufferWriter 类） 178
  9.3.6　PrintWriter（PrintWriter 类） 181
 9.4　File class（文件类） 181
 Exercises 184

Chapter 10　Multi-threading（多线程） 186

 10.1　Concept of multi-threading（多线程的概念） 186
  10.1.1　What's a thread?（什么是线程？） 186
  10.1.2　Thread vs process（线程与进程） 186
 10.2　Life cycle of a thread（线程的生命周期） 188

10.3　Creating threads（创建线程）..190
　　10.3.1　Direct approach of creating a thread（直接法创建线程）..............190
　　10.3.2　Indirect approach of creating a thread（间接法创建线程）............192
10.4　Main thread（主线程）...194
10.5　Methods of Thread class（线程类的方法）...196
10.6　Thread synchronization（线程同步）..199
10.7　Communication between threads（线程间的通信）..............................202
Exercises...203

Chapter 11　Networking（网络）..205

11.1　Concept of networking（网络的概念）..205
11.2　URL class（URL 类）...206
11.3　Sockets communication（套接字通信）...209
　　11.3.1　How Socket communication works?（套接字通信是如何进行的？）...........209
　　11.3.2　Ports（端口）..210
11.4　ServerSocket and Socket classes（ServerSocket 和 Socket 类）..............210
　　11.4.1　Tasks of each side（双方的任务）...211
　　11.4.2　Data transmission（数据传输）..212
11.5　Serving multiple clients（服务多个客户）..217
Exercises...219

References..222

Chapter 1
Introduction(简介)

Java is a programing language. Sun released Java in 1995. After years developing, nowadays, Java plays more important rule in Internet, gameplay and mobile communication programming. Java is based on C++, is a "pure" object-oriented programing language. Java inherits the class concept of C++ and adds multi-thread, exception dealing and other advanced techniques, meanwhile abandoned the pointer operator of C++ language.

Java 是一种程序设计语言,Sun 公司于 1995 年首次发布,经过多年发展,今天 Java 在互联网、游戏、移动通信等领域发挥着重要的作用。Java 在 C++ 的基础上,成为"完全"的面向对象的程序设计语言。它继承了 C++ 类的概念,增加了线程、异常处理等先进技术,同时摒弃了 C++ 语言的"指针"。

1.1 Characteristics of Java(Java 的特点)

The main characteristics of Java are as below:
- Simple
- Object-oriented
- Robust
- Sophisticated compiler
- Runtime exception-handling
- Secure
- Multi-thread
- Cross-platform

There are three different versions of Java:
- Java SE (Standard Edition)
- Java EE (Enterprise Edition)
- Java ME (Miniature Edition)

Java SE is the standard version of Java, which is used in this book. Java EE is the enterprise version of Java. Java ME is the miniature version of Java, which is applied to embedded, mobile phone and other mobile communication applications.

Java SE 是 Java 的标准版，本书使用这个版本。Java EE 是企业版，Java ME 是微型版，用于嵌入式、手机和移动通信方面的应用开发。

1.2 Developing tools（开发工具）

In order to design a Java program, some developing tools are needed. Java is an interpretive language, so the developing tools should be adapted to it.

1.2.1 Interpreted language vs compiled language（解释性语言与编译性语言）

To run Java programs, the source files are compiled to produce bytecode files. The bytecode is machine-independent. To run bytecode files, a Java interpreter, which is part of the Java Virtual Machine (JVM) is needed.

To make it clear, what "interpretive" means, let's go back to C++ language. C++ is a compiled language, all codes are once compiled into a native machine language before running. Suppose a C++ source file mycpp.cpp is provided, which is compiled to produce an object file, a machine-dependent (machine language) file, mycpp.obj. Linking mycpp.obj with some library object files, an executable file mycpp.exe is produced, which can only run on a certain kind of Operating System where it is compiled and linked. The converting procedure of a C++ program, from a source file to an executable file is as shown in Figure 1-1.

运行 Java 程序前，需要将源程序编译成字节码文件。字节码文件与机器无关，由 Java 虚拟机里的解释程序解释后执行。

为了弄清楚什么是"解释性"，让我们回到 C++ 语言。C++ 是编译性语言，运行前，对源程序进行编译，生成机器语言程序。假设 C++ 源程序是 mycpp.cpp，编译后生成与机器相关的目标文件为 mycpp.obj 文件，将 mycpp.obj 与库文件连接，生成可执行文件 mycpp.exe。这个可执行文件只能运行在对源文件进行编译和连接的那类机器上。源文件到可执行文件的转换过程如图 1-1 所示。

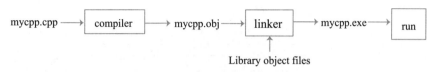

Figure 1-1 Converting of C++ files

On the other hand, a Java source file is compiled to produce bytecode (.class) files only, which can't run directly on an Operating System.

Suppose a Java source file myjava.java is provided, which is compiled to produce one or several bytecode files, but unlike C++ executable files, they can't run directly on an Operating System. The bytecode files are binary code files, which must be interpreted by a JVM, that is, they rely on a JVM. A Java Virtual Machine explains bytecode files and executes the explained codes. With bytecode files, Java realizes the cross-platform characteristic, "compilation once, execution everywhere". Bytecode files can run on different JVMs, Linux JVM, Windows JVM, …, Unix JVM.

The converting procedure of Java program files is shown in Figure 1-2.

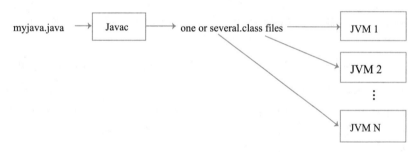

Figure 1-2 Converting of Java files

Java 源程序仅仅被编译成字节码文件，即一个或多个 .class 文件。字节码文件是二进制文件，不能直接在某个操作系统上运行，它必须由 Java 虚拟机 JVM 解释并运行。只要有 JVM，字节码文件可以在任何类型的机器上运行，具有一次编译、处处运行的跨平台特性。

1.2.2 Java Virtual Machine（虚拟机）

A Java virtual machine (JVM) is a software computing machine that enables a computer to run a Java program. The architecture of a JVM is shown in Figure 1-3. There are four main units in it.

- Class Loader
- Bytecode Verifier
- Interpreter
- Just-in-time compiler (JIT)

JVM 是 Java 虚拟机，是用软件实现的处理器，用于运行 Java 程序，其组成如图 1-3 所示，它包含 4 个主要模块。

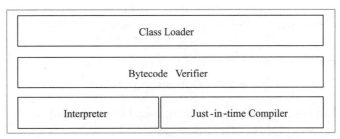

Figure 1-3　Structure of Java virtual machine

A bytecode (.class) file contains binary codes of a Java class. The class loader recognizes and loads Java .class files to memory. The bytecode verifier verifies all bytecode before it is interpreted and executed.

For different machines, different JVMs are needed. As long as a corresponding JVM is installed on a computer, the computer can run any Java bytecode programs.

一个字节码文件对应一个 Java 类的二进制代码，它由 JVM 的类装载器（Class Loader）装入指定的存储空间，并被字节码检验器（Java Verifier）检查，再由 JVM 的解释器（Interpreter）解释并执行。

不同类型的计算机需要不同的 JVM。只要计算机安装了对应的 JVM，就可以运行任意 Java 字节码程序。

Java bytecode executes while interpreting, the execution will always be slower than the execution of the codes that are compiled into a native machine language. This problem is solved by the just-in-time (JIT) compiler. A JIT compiler can translate Java bytecode into a native machine language while executing the program. For some frequently used codes, the translated parts of the program can be executed much more quickly.

1.2.3　Java Runtime Environment（运行环境）

Java Runtime Environment (JRE) is a software package, which contains the Java virtual machine and the Java class library.

Java 运行环境是个软件包，包含 JVM 和 Java 类库。

1.2.4　Java Development Kit（开发工具）

If you want to write programs, you need to prepare Java Development Kit (JDK). JDK contains Java Runtime Environment (JRE) and developing tools: Java compiler "Javac" for compiling a Java source file, Java interpreter "Java" for explaining and running Java bytecode. JDK is a superset of a JRE. It is free of charge. The latest version 1.8 of JDK is available at http://www.oracle.com/.

JDK 是开发 Java 程序的开发工具，包含 Java 运行环境 JRE，Java 编译器"Javac"和 Java 解释器"Java"。

1.3 A simple Java program（一个简单的Java程序）

【Example 1-1】HelloWorld.java

```
class HelloWorld {
  public static void main(String args[]) {
    System.out.println("Hello World!");
  }
}
```

The function of this program is to print "Hello World!". A Java program is composed of one or more classes, at least one class. This program contains a class "HelloWorld", which has a method "main". The main is the start point to execute the Java program, which is the same as C++ program. A class that contains the method main is referred to as the main class. A class can have several methods.

A method is a collection of statements, just like a function of C or C++ program. In this example, there is only one statement in method main, which outputs "Hello World!" to the screen.

这个程序的功能是在显示器上输出"Hello World!"。Java程序由一个或多个类组成，至少一个类。这个程序包含一个"HelloWorld"类，类里有个称为main的方法。与C++类似，main是程序执行的入口，包含main方法的类是主类，类可以包含很多方法。

方法是语句的集合，与C或C++的函数类似。此例中，main中只有一条输出"Hello World!"的语句。

1.3.1 How to run a Java program?（如何运行Java程序？）

Figure 1-4 shows the steps of a Java source program how is compiled, is executed and outputs "Hello World!" to the screen.

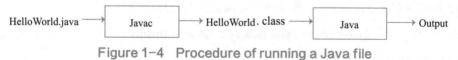

Figure 1-4　Procedure of running a Java file

Suppose you are in a Microsoft Windows Operating System, JDK is installed at your computer successfully, and the source file is at the root of disk f. You need to go into command window, CMD Window first. When you are there, the screen looks like:

```
C:\Users\Administrator>_
```

Type in command "f:" at the cursor and press Enter key, you enter into the root of disk f.

```
F:\>_
```

Type in "javac HelloWorld.java", calls the compiler to compile the source file HelloWorld.java, for instance:

```
F:\>javac HelloWorld.java
```

If there are no errors in the HelloWorld.java, then a bytecode file HelloWorld.class is produced. To run the HelloWorld.class, you need to call interpreter of the JVM by typing command java.

```
F:\>java HelloWorld
```

"Hello World!" is seen.

1.3.2 Rules of naming a Java source file（Java源文件的命名规则）

A Java source file is consists of one or more classes. The file name must follow some rules. File naming rules:

(1) The file name must be the same as the class name if only one class declaration exists.

(2) At most, only one class is public and the file name must be the name of the public class. If more than one classes need to be public, the source file should be split into several source files.

(3) Without the main class (a class that contains the main method is called the main class), if there are several classes and none is public in a file, the file name can be anyone of the class names.

(4) If a source file contains the main class, the file name must be the same as the main class.

Java源文件包含一个或多个类，文件的命名必须遵循特定的规则。文件的命名规则如下：

（1）如果源文件里只有一个类，文件名必须是该类的类名。

（2）每个源文件里至多只能有一个public类，并且文件名必须是该public类的名字。如果一个源文件里的多个类都需要设成public，要将这个源文件分解成多个源文件。

（3）在没有主类的情况下（含有main方法的类称为主类），如果源文件里有多个类，并且每个类都不是public，文件名可以是其中任何一个类的类名。

（4）如果源文件里包含主类，则文件名必须是主类的类名。

【Example 1-2】 Two classes exist in a file, none is public

```
class First {
  static String Message="Hello Java!";
}
class Second {
  public void hello() {
    System.out.println(First.Message);
  }
}
```

The file name can be First.java or Second.java. After successfully compiled, two bytecode files: First.class and Second.class are produced. But without the main class, they can't run alone.

【Example 1-3】 **Two non-public classes exist in a file, one is the main class**

```
class First {
  static String message="Hello Java!";
}
class Second {
  public static void main(String args[]) {
    System.out.println(First.message);
  }
}
```

The file name must be Second.java, because the main method in it. Class Second is the main class.

【Example 1-4】 **Two public classes needed, a file must be split into two**

```
public class First { //First.java
  static String message="Hello Java!";
}
```

【Example 1-5】 **Two public classes needed, a file must be split into two**

```
public class Second { //Second.java
  public static void main(String args[]) {
    System.out.println(First.message);
  }
}
```

Notice that the number of .class files is independent of the number of source files, but the number of classes. Example 1-3 produces two .class files, example 1-4 and example 1-5 totally produce two .class files, First.class and Second.class.

1.3.3 Rules of naming classes, variables and methods （类、变量和方法的命名规则）

Choose meaningful names.

1. Classes

Capitalize the first letter of each word in the name, for instance ComputeArea.

2. Variables and methods

Use lowercase. If the name consists of several words, concatenate all in one, use lowercase for the first word, and capitalize the first letter of each subsequent word in the name, for instance computeArea.

3. Constants

Capitalize all letters in constants, and use underscores to connect words, for instance PI, MAX_VALUE.

名字要体现用意。

1. 类名

大写每个单词的第一个字母，例如，ComputeArea。

2. 变量和方法名

用小写字母。如果名字包含多个单词，小写第一个单词，大写其后的每个单词的第一个字母，例如，ComputeArea。

3. 常数名

大写所有字母，用下划线连接不同单词，例如，PI，MAX_VALUE。

Exercises

1. Install JDK and configure environment variables.
2. Write, compile and run program HelloWorld.java.

Chapter 2
Java Foundation（Java 基础）

This chapter will introduce you the basic concepts of Java.

Suppose all students have learnt C or C++ programming language, so some subjects in C or C++ will not be repeated or described with fewer words, so as to we can focus on new and different subjects of Java.

2.1 Primitive data types（基本数据类型）

Java has two kinds of data types, reference and primitive. Primitive data types are provided by Java program language. Java determines the size of a variable of the primitive data type. There are eight primitive data types in Java (see Table 2-1).

1. Integer
(1) byte (1 byte).
(2) short (2 bytes).
(3) int (4 bytes).
(4) long (8 bytes).

2. Floating point
(1) float (4 bytes).
(2) double (8 bytes).

3. Textual
char (2 bytes).

4. Logical

boolean (1 byte).

The size of a primitive type variable remains the same on all platforms (standardized). All the primitive data types and their sizes are listed in Table 2-1.

Table 2-1 Primitive data types

Primitive data type	Size(bits/bytes)	Minimum	Maximum
byte	8/1	-128	127
short	16/2	-2^{15}	$2^{15}-1$
int	32/4	-2^{31}	$2^{31}-1$
long	64/8	-2^{63}	$2^{63}-1$
float	32/4	$\pm 3.4 \times 10^{-38}$	$\pm 3.4 \times 10^{38}$
double	64/8	$\pm 1.7 \times 10^{-308}$	$\pm 1.7976931348623157 \times 10^{308}$
char	16/2	\u0000	\uFFFF
boolean	8/1	—	—

2.1.1 Integer and floating point types（整型与浮点型）

All numeric data types, integer and floating point, are signed. The integer type in C++ is divided into signed and unsigned. Integer constants are assumed to be int type by default, like 100, the default type can be converted to long, by attaching L or l.

e.g. 100　　int

　　100L　long

数值型变量，整型和浮点型都是有符号数。C++ 中，整型分有符号和无符号整型。整型常量的默认类型是 int 型，在常量后面加 L 或者 l 可以将其变成 long 型。

For instance:

5/2 yields an integer.

5.0/2 yields a double.

5%2 yields an integer (remainder of the division).

(6+10)%7 yields an integer.

Floating point constants are assumed to be double type by default. The default type can be converted to float by attaching F or f.

浮点型常量的默认类型是 double，在常量后面加 F 或者 f 可以将其变成 float 型。

For instance:

```
float k=1.2;
```

Error, a double constant can't be assigned to a float variable.

```
float k=1.2f;
```

Right, a float constant is assigned to a float variable.

```
float k=(float)1.2;
```

Right, a double constant is casted to float before assigned to a float variable.

2.1.2　Boolean type（布尔型）

There are two boolean cnstants (true and false) can be assigned to a boolean variable, true≠1 and false≠0.

有两个布尔常量 true 和 false 可以为布尔变量赋值，true ≠ 1，false ≠ 0。

【Example 2-1】**Boolean type**

```
class BoolType {
  public static void main(String args[]) {
    boolean b=false;
    System.out.println("b is "+b);
    b=true;
    System.out.println("b is "+b);
    System.out.println("10>9 is "+(10>9));
  }
}
```

Output:

b is false
b is true
10>9 is true

2.1.3　Character type（字符型）

Prior to Java, most programming languages in common use the ASCII character set, 8-bit encoding. ASCII character set provides 256 different characters. Consider all of the human languages in the world and the various symbols they use, it turns out that 256 characters are not enough. Therefore, Java adopts the Unicode character set.

Unicode, a 16-bit encoding scheme, supports written texts in the world's diverse languages, such as Chinese, Japanese. A Unicode character takes 2 bytes, so Unicode can represent 65, 536 characters. The ASCII character set is the subset of the Unicode from \u0000 to \u007f (the first 128 characters). A character constant is denoted using single quotes ' '.

Java 之前，很多程序设计语言都采用 ASCII 字符集，8 位编码。ASCII 字符集提供 256 个不同的字符。然而，对于世界上各种各样的语言，各种各样的字符，256 个是远远不够的，因此，Java 采用 Unicode 字符集。

Unicode，16 位编码，支持包括中文、日文的各种各样语言的字符。一个 Unicode 字符需要 2 个字节，共有 65 536 个字符。ASCII 字符集是 Unicode 字符集的子集，即 Unicode 的前 128 个字符。字符常量用一对单引号括起来。

For instance:

```
char ch;
ch='良';
```

A Chinese character is assigned to ch.

```
ch='k';
```

Letter k is assigned to ch.

【Example 2-2】**Character type**

```
class Test {
  public static void main(String args[]) {
    char a;
    a='良';
    switch(a) {
      case '优': System.out.println(90+"分以上"); break;
      case '良': System.out.println(90+"分以下"); break;
      case 'f' : System.out.println(60+"分以下");
    }
  }
}
```

Java provides a set of escaped characters. They are listed in Table 2-2.

Table 2-2　Escaped characters

Description	Escaped character	Unicode
Backspace	\b	\u0008
Tab	\t	\u0009
New line	\n	\u000A
Carriage return	\r	\u000D
Back slash	\\	\u005C
Single quote	\'	\u0027
Double quote	\"	\u0022
Octal value	\ddd	
Hexa decimal	\uxxxx	

For instance:

```
System.out.println("\u0041\n\u0061\t\141\n");
```

or

```
System.out.println("A\na\ta\n");
```

 Output:

A

a　*a*

2.2　Reference types（引用型）

Basically, anything that is not a primitive (an int, a float, etc.) is a reference. The notion of a reference is similar to the pointer in C++.

Suppose you have defined a class called MyClass,

```
MyClass myr=new MyClass();
```

Where, a MyClass type variable myr, a reference, is defined, and an object of MyClass type is created. The object acquired memory and assigned its storage address to myr.

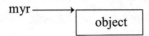

Reference myr does not contain an object of MyClass. Rather, it contains the memory address where that object really resiclents. The reference is very important in Java. There are three distinct references: class, interface and array.

For a primitive type variable, such as int k, you can assign an int type variable or an integer constant to k. However, the only constant can be assigned to a reference is "null". This special value is known as the null reference. The null reference has no type at runtime but may be casted to any reference type.

除基本数据类型外都是引用类型。引用类型与 C++ 的指针类似。假设用户已经定义了一个 MyClass 类：语句 MyClass myr=new MyClass(); 定义了一个 MyClass 类型的变量 myr，myr 就是引用，new 创建了一个 MyClass 类型的对象并将对象的地址赋给了 myr。

引用 myr 并不包含 MyClass 类的对象，引用的值是对象的存储地址。引用在 Java 中很重要，有三种不同的引用：类、接口和数组。

对于基本数据类型，例如 int k，你可以用一个整型变量或者一个整型常数给它赋值。而对于引用，null 是唯一可以给引用赋值的常数。程序运行时，null 无类型，但可以转换成任意引用类型。

2.2.1　A class is a data type（类是数据类型）

A class is a data type to describe a set of objects that have identical characteristics (data members) and behaviors (member methods). A class is really a data type, an abstraction of objects. Once a class is established, you can create as many objects of that class as you like.

Virtually, all object-oriented programming languages use "class" keyword. A class is a user defined data type.

class 是数据类型，用于描述一组具有相同特性（数据成员）和行为（成员方法）的对象，是对同类对象特征的抽象。一旦定义了一个 class 类型，用户可以生成多个该类型的对象。

事实上，所有面向对象的程序设计语言都使用 class 关键词来表示类。class 是用户定义的数据类型。

For instance:

```
class MyClass {
  /*class body*/
}
```

This introduces a data type, you can create objects of this type using new.

```
MyClass obj=new MyClass();
```

When you define a class, you can put two types of elements in your class: fields (member variables or data members) and methods. A field is an object of any type or a primitive type variable. Each object has its own storage for its fields.

类里包含 fields（成员变量，数据成员）和方法成员。成员变量可以是基本数据类型也可以是任意类型的对象。每个用 new 创建的 class 类型的对象，都有自己的存储空间存放其 fields。

【Example 2-3】**Member variables**

```
class DataOnly {
  int i=9;
  double d;
  boolean b;
}
```

This class doesn't do anything except declares data. You can create an object to get a piece of memory to hold its three data members, for instance:

```
DataOnly data1=new DataOnly();
DataOnly data2=new DataOnly();
new DataOnly();
```

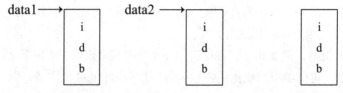

Losing its address, the third object can never be referenced to, and the storage soon will be reclaimed by the Java garbage collection mechanism.

2.2.2　A class type variable is a reference（类类型的变量是引用）

A class type variable is a reference, you can assign it with an object. Java manipulates objects with references. Remember, in Java, an object (an instance) of a class is created using new and new brings back the memory address of the object.

一个 class 类型的变量是引用，用户可以将一个新建的对象与之关联。Java 通过引用操作对象。Java 中用 new 来创建对象（实例），new 返回对象的地址。

For instance:

Suppose Student is a class.

Student sc, sb;

```
sc=new Student(101,"Jeff Bissle");
sc and sb are references.
```

However, in C++, the statement Student sc(101,"Jeff Bissle"); is defining an object or instance and can't split it into two parts.

Once an object is assigned to sc, by sc, members of the object can be referred to.

```
sc=new Student(101,"Jeff Bissle");
```

Object (instance) of a class is created using new. Suppose outa is a method of class Student.

```
sc.outa();
sb=new Student(102,"Lisa Urbania");
sb.outa();
```

Examples of references defining, suppose MyObject is a class.

```
MyObject obj;
obj=new MyObject();
```

The value of obj is the address of the newly created object using new.

```
MyObject ob=new MyObject();
MyObject anotherOb=null;
```

You can't assign a constant 0 to a reference, for instance:

```
MyObject anotherOb=0; //error
```

【Example 2-4】**A reference variable**

```
class NULL {
  public static void main(String args[]) {
    NULL c=null; //c is a reference variable
    System.out.println("c value is:"+c);
    c=new NULL(); //assign address of the object to c
    System.out.println("c value is:"+c);
  }
}
```

Output:

c value is: null

c value is: NULL@1c5f743

In method main, c must be assigned a valid value, before referring to. You can't let it as such:

```
NULL c;
System.out.println("c value is:"+c);
```

Because c is a local reference variable in the main method and can't be used before assigning. In line 2 of the output, the Hash address of the object (classname+@+Hash address) is displayed.

在这个 main 方法中, 给 c 赋值后才能访问它。不能像下面这样:

```
NULL c;
System.out.println("c value is: "+c);
```

因为 c 是主方法里的局部引用, 赋值前不能访问。输出的第二行包含了对象的 Hash 地址, 它由类名 +@+Hash 地址组成。

2.2.3 Interface type（接口类型）

An interface acts as a template. An interface cannot be instantiated. A class can implement multiple interfaces and be referred to by any of the interfaces that it implements.

The interface type will be introduced at later chapter of this book, which is a type not available in C++.

2.3 Identifiers（标识符）

An identifier is the name of a class, a method, a variable or a constant. An identifier is any sequence of uppercase letters, lowercase letters, digits, $ (dollar sign) and _ (underscore).

An identifier should follow rules: Cannot start with a digit, cannot be a reserved word, case-sensitive, S and s are two different identifiers

标识符是类、方法、变量或常量的名字。标识符由字母、数字、下划线和美元符组成。标识符不能以数字开头，不能使用保留字，区分大小写。

For instance:

```
int 我是 823;
```

Define an int variable, the variable name is 我是 823. Characters that form an identifier can come from diverse languages.

Examples of valid identifiers:
timeOfDay, 你好 , $ 美元 , temp_val, a3, $_, _abc, $$hello$.
Examples of invalid identifiers:
3more, x+y, 3$, #foo, @2, 123, oe**name, 2floor.

2.4 Default values of fields（成员变量的默认值）

Fields (member variables), primitive types or reference types, have default values for all class objects. Local variables and references within methods have no default values. When a method is called, a variable or a reference defined in the method is allocated with a piece of memory, the size of the storage depending on its type, int, double or class. The local variables have no initial values, so you can't use them before assignment. All fields, defined in classes and outside methods have default values.

类的所有成员变量、基本数据类型或引用类型，都有默认值。在方法里定义的变量称为局部变量，局部变量没有默认值。调用方法时，在方法里定义的变量或者引用获得存储空间，存储空间的大小与变量的类型有关。方法里的局部变量没有默认值，因此赋值前，不能使用它们。在类内，方法外定义的成员变量都有默认值。

A field is guaranteed to get a default value if you do not initialize it. The default values of fields are listed in Table 2-3.

Table 2-3   Default values of fields

Type	Default value
boolean	false
char	'\u0000'
byte	(byte)0
short	(short)0
int	0
long	0L
float	0.0f
double	0.0
reference	null

Pay attention to the last line, the reference (class, interface and array) type, its default value is null. You can use any fields if needed without initialization.

【Example 2-5】**Default values of fields**

```
class Default {
  private int value;
  public int getvalue() {
    return value;                //return default
  }
  public static void main(String args[]) {
    Default c;                   //local variable, no default
    System.out.println("c value is: "+c); //error
    c=new Default();
    System.out.println("c value is: "+c); //okay, c holds a value
    System.out.println("c value is: "+c.getvalue());
    int a;                       //no default
    System.out.println("a value is: "+a); //error
  }
}
```

【Example 2-6】**Default values of fields**

```
class Default {
  private int value;
  public int getvalue() {
    return value;
  }
  public static void main(String args[]) {
    Default c=new Default(); //c no default, but is assigned a value,
                             //address of the new object
    System.out.println("value="+c.getvalue());
    InitValue d=new InitValue();
    System.out.println("score="+d.getFloat());
    System.out.println("reference="+d.getReference());
    System.out.println("ch="+d.getChar()); // '\u0000', invisible character
  }
}
```

```
class InitValue {                          //member variables
  private Default ref;                     //ref is a reference
  private float score;
  private char ch;
  float getFloat() {return score;}
  Default getReference() {return ref;}
  char getChar() {return ch;}
}
```

Output:

value=0

score=0.0

reference=null

ch=

The first line in class InitValue, defines a reference, its default value is null.

2.5 Where data store?（数据存储在何处？）

While a program is running, how memory is arranged? There are 3 different places to store data and codes.

1. Stack

All "automatic" variables (local variables defined in methods) are on the stacks, so it's much more efficient. Stacks lives in the RAM area.

2. Heap

This is a general-purpose pool of memory (also in the RAM area) where all Java objects live. Unlike the stack, the compiler doesn't need to know how long that storage must stay on the heap. There's a great of flexibility in using storage on the heap.

3. Constant

Constant values are often placed directly in the program code, which is safe since they can never change.

程序运行时，内存是如何分配的？有 3 个不同的区域存储数据和代码：

1. 堆栈

所有局部变量（自动变量）都在堆栈里，堆栈在 RAM 里。

2. 堆

所有对象都在堆里，堆也在 RAM 里。堆在使用上比堆栈更为灵活方便。

3. 常数

通常直接包含在程序代码里，常数不能被修改。

2.6 Operators（运算符）

In this section, only the operators that are different from the ones in C++ and the ones introduced few in C++ are reiterated.

2.6.1 Arithmetic operators（算术运算符）

+、-、*、/、% are operators both in C++ and Java. Only % is little different.

The modulus operator, %, returns the remainder of a division operation. It can be applied to floating-point types as well as integer types (in C++, only applies to integer types).

【Example 2-7】**Arthmetic operators**

```java
class Modulus {
  public static void main(String args[]) {
    int x=42;
    double y=42.3;
    System.out.println("x mod 10="+x%10);
    System.out.println("y mod 10="+y%10);
  }
}
```

Output:

x mod 10 = 2

y mod 10 = 2.299999999999997

2.6.2 Logical operators（逻辑运算符）

Operator	Name
!	NOT（逻辑非）
&&	AND（逻辑与）
||	OR（逻辑或）
^	XOR（逻辑异或）

There are four logical operators, NOT, AND, OR and XOR. Some of you may not so familiar with XOR, Table 2-4 is the truth table of XOR.

Table 2-4 Truth table of XOR(^)

p1	p2	p1^p2
false	false	false
false	true	true
true	false	true
true	true	false

A logic operating result can only be true or false, different from 1 or 0 of C++.

【Example 2-8】**Logical operators**

```
class Example {
  public static void main(String args[]) {
    boolean b;
    b=(2>3)&&(3<2); //where (3<2) won't be checked further
    System.out.println("b="+b);
    b=false||true;
    System.out.println("b="+b);
  }
}
```

2.6.3 Bitwise operators（位运算符）

The bitwise operators are low-level operators that manipulate the individual bits that make up an integer value. The bitwise operators are most commonly used for testing and setting individual flag bits.

&, | and ^ operators are different from the logical operators &&, || and ^.

~ is the bitwise NOT operator. It's different from the logic NOT operator "!". With the logic "!", !true comes to false. With the bitwise NOT "~", it inverts each bit of its single operand, converting ones to zeros and zeros to ones.

For instance:

(1) Bitwise AND (&) 按位与运算。

```
byte a=10;
byte b=7;
```

The result of 10&7 is 2.

  00001010
&00000111
  00000010

(2) Bitwise OR (|) 按位或运算。

```
byte a=10;
byte b=7;
```

The result of 10|7 is 15.

  00001010
|00000111
  00001111

(3) Bitwise XOR (^) 按位异或运算。

^ is exclusive OR.

```
class XOR {
  public static void main(String[] a) {
    System.out.println(true^(9>2));
    System.out.println(10^7);
```

```
    }
}
```
The result of 10^7 is 13.
  00001010
^ 00000111
  00001101

(4) Bitwise NOT 按位非运算。

```
byte b=~12;
```

~00001100, the value of b is 13 (11110011)

Remember that data in memory are all the complements, sometimes the inverting results may not what you expected.

2.6.4 Left shift (<<) and signed right shift (>>) operators（移位运算符）

The << operator shifts the bits of the left operand left by the number of the right operand. High-order bits of the left operand are lost, and zeros are shifted in from the right. If the left operand is positive and the right operand is n, the << operator is equivalent to multiplying that number by 2^n (nth power of 2).

The >> operator shifts the bits of the left operand right by the number of the right operand. If the left operand is positive and the right operand is n, the >> operator is the same as integer division by 2^n and zeros are shifted to the higherder bits, or else if the left operand is negative, the ones are shifted to the highorder bits.

10>>1   //00001010>>1=0000 0101
27>>3   //00011011>>3=00000011
–50>>2   //1100 1110>>2=11110011

【Example 2-9】 << and >> Operators

```
class Shift {
  public static void main(String args[]) {
    int a,b;
    a=5;
    b=a<<4; //left shift by 4 bits
    System.out.println(b);
    b=a>>1; //right shift by 1 bit
    System.out.println(b);
    b=~a|a; //Not then OR
    System.out.println(b);
    b=a&0x0ff; //AND
    System.out.println(b);
    b=a^a; //Exclusive OR
```

```
        System.out.println(b);
        boolean c=true;
        System.out.println(c^(b==0));
    }
}
```

Output:

80
2
–1
5
0
False

2.6.5 Assignment and conditional operators（赋值运算符和条件运算符）

The assignment operators are right-associative (右结合性), which means that the assignments a=b=c are performed right-to-left, as a=(b=c).

The conditional operator ?: allows to embed a condition within an expression. There are three operands in it. The first operand must evaluate to a boolean value. The second and third operands can be of any type, but they must be the same type.

expression1? expression 2: expression 3

For example:

```
int max=(x>y) ? x:y;
String name=(name !=null) ? name: "unknown";
```

2.6.6 String operator "+" and "+="（字符串运算符 "+" 和 "+="）

The "+" and "+=" operators are not only for numerical operation, but also for concatenating strings. If a String variable or String constant "+" or "+=" other types of variables or constants, all they will be casted to String type strings first (remember that the compiler automatically turns a double-quoted sequence of characters into a string).

"+" 和 "+=" 运算符不仅用于数值运算也可以用于字符串的连接。如果一个 String 变量或者 String 常数 "+" 或者 "+=" 其他类型的变量或者常数，所有其他类型都先被转成 String 字符串。

```
int x=0, y=1, z=2;
String s="x, y, z";
s=s+x+y+z;
System.out.print(s + (x + y + z)); //x, y, z 0123
```

```
String s="x, y, z";
s=s+(100>2);
System.out.print(s); //x, y, z true
```

【Example 2-10】String operator

```
public class StrOp {
  public static void main(String[] args) {
    float x=11;
    System.out.println(x+''); //Unicode of blank, 32
    System.out.println(x+" ");
  }
}
```

Output:

43.0

11.0

2.6.7　Special operators（特殊运算符）

There are three special operators in Java:

(1) .　　Object member access

(2) [ ]　Array element access

(3) new　Object creation

For instance:

```
System.out.println("Hello");
```

In Java, objects are created with the new operator.

2.7　Casting（类型转换）

Java allows you to cast any numerical primitive type to any other numerical primitive type. Casting is an explicit or implicit modification of a variable's type.

Casting allows us to view data as a different type than it was given. Java will automatically change one type of data into another. If you assign an integral value to a floating point variable, the compiler will automatically convert the int type to the float type.

2.7.1　Widening and narrowing（拓宽与缩窄）

There are two types of conversions that can be done to data, widening and narrowing conversions. Widening conversions can be done implicitly (by the compiler) and narrowing conversions must be done explicitly.

A widening conversion is when you take a variable and assign it with a value of smaller type.

A narrowing conversion is when you take a variable and assign it with a value of larger type.

There are two types of conversions: automation and coercion. The automatic conversion is as shown in Table 2-5.

Table 2-5 Automatic conversion

one	another	result
byte,short,char	int	int
byte,short,char,int	long	long
byte,short,char,int,long	float	float
byte,short,char,int,long,float	double	double

1. Automatic conversion: widening

(1) Lifting from small to large

(2) Basic values: scope of ranges

byte→int, char→int, long→float, float→double

2. Coercion: narrowing

(1) Compressing from large to small

(2) Syntax: (type) value

double→long, float→int, double→float

Here are some examples of widening and narrowing.

```
float g=(float)1.2;        //a narrowing conversion
double f=g;                //widening, cast not really required
g=(float)(f+g);
```

Widening:

```
byte x=100;                //1 byte
short y;                   //2 bytes
y=x;
```

Narrowing:

```
byte x;                    //1 byte
short y=1000;              //2 bytes
x=y;                       //error, no enough to storage y
x=(byte)y                  //right
```

Examples of error assignments:

```
byte b=1000;               //cannot be hold by a byte variable
double k=5.0;
int s=k;                   //error, but right in C++
s=(int)k;                  //right
```

While coercion, narrowing, truncation may happen.

```
double vd=0.7;
float vf=0.7f;
System.out.println((int)vd);
```

```
The output is 0.

short s=130;
byte b=(byte)s;
value of b is - 126

int i='\uffff'; //65535
short s=(short)'\uffff'; //-1
```

2.7.2 Char, byte and short produce int results（Char, byte 和 short 转换为 int 型）

You will see the effects of promotion with the arithmetic operators. Each arithmetic operation on any of char, byte and short types produces an int result. They are lifted as int values before calculation. So byte and short types are used for storage, not for calculation.

```
int i='a';         //same as int i=(int)'a';
char c=97;         //same as char c=(char)97;

byte a=40, b=50, c=100;
int d=a*b/c;
```

a, b, c is lifted as int before * and /

```
c=a*b;
```

A compiling error occurs because a*b is lifted as int.

```
Let c=(byte)a*b; or c=(byte)(a*b);
```

Which can cast int to byte?

2.8 Flowing control（流程控制）

A Java program runs sequentially, one statement is interpreted and ran after the other. The sequential procedure may change on conditions of control statements. There are two types of control statements.

1. Conditionals

(1) if, if-else statements

(2) switch statement

2. Loops

(1) for

(2) while

(3) do-while

The control condition must be true or false.

For instance:

```
if(a>b) {…}
if(a>'a'&& a<'z') {…}
if(true) {…}
while(i<=100) {…}
while(false) {…}
Can't be if(2) {…}
```

2.8.1　Basic controlling statements（基本控制语句）

(1) if

(2) if-else

(3) switch

(4) while

(5) do-while

(6) for

(7) return

(8) break

(9) continue

(10) goto

【Example 2-11】Using break and continue statements

```
public class For {
  public static void main(String[] args) {
    for(int i=0; i<100; i++) {
      if(i==74) break;         //out of for loop
      if(i%9!=0) continue;     //next iteration
      System.out.print(i+" ");
    }
  }
}
```

Output:

0 9 18 27 36 45 54 63 72

2.8.2　Foreach statement（foreach 语句）

Java foreach loop iterates through arrays and collections. It's the enhanced form of for loop that enables us specifying an array or other collections working with their elements. The enhanced for loop of Java works just alike the way of loop. The current element can be assigned to a variable inside for loop. You may perform a certain action with that element and to the next item until all elements are entertained.

Java foreach 循环遍历数组和集合。它是增强型的 for 循环语句，可以用它访问数组或集合的所有元素。它的循环过程与 for 相同，将数组或集合的当前元素赋值给一个变量，取这个变量的值或者让它参与表达式的计算，接着对下一个元素重复相同操作，直到访问完所有元素。

In example 2-12, the regular "for" statement is adopted to display default values of each element of an array.

【Example 2-12】**For statement**

```
public class For {
  public static void main(String[] args) {
    float f[]=new float[10];
    for(int i=0;i<10;i++)
      System.out.print(f[i]+" ");
  }
}
```

In example 2-13, the foreach loop is adopted to display default values of each element of an array. You don't have to define an int variable to count through a sequence of items, foreach takes each item automatically.

例 2-13 利用 foreach 语句显示数组元素的默认值。不需要像 for 语句那样定义一个 int 型的循环变量，foreach 语句自动取出数组的每个元素。

【Example 2-13】**Foreach statement**

```
public class ForEach {
  public static void main(String[] args) {
    float f[]=new float[10]; //array with 10 float items
    for(float x: f)
      System.out.print(x+" ");
  }
}
```

【Example 2-14】**Foreach statement**

```
public class ForEachString {
  public static void main(String[] args) {
    for(char c: "An African Swallow".toCharArray())
      System.out.print(c + " ");
  }
}
```

Output:

AnAfricanSwallow

【Example 2-15】**Foreach statement**

```
public class Main {
  public static void main(String[] args) {
    for(String s:args)
      System.out.print(s+" ");
  }
}
```

2.9 Arrays（数组）

An array is a group of variables of the same data type.

2.9.1 Define arrays（定义数组）

```
int[] a=new int[20];
```

or

```
int a[]=new int[20];
A s[]=new A[12];
String[] str=new String[30];
```

【Example 2-16】**Define arrays**

```
public class Test {
  public static void main(String[] ss) {
    int[] a=new int[20];
    A s[]=new A[12];
    String[] str=new String[30];
    System.out.println("int:"+a[0]+"_A:"+s[11]+"_String:"+str[20]);
  }
}
class A {}
```

Output:

int:0_A:null_String:null

2.9.2 Initialize arrays（初始化数组）

All array elements have default values according to data types of arrays. The defaults can be changed while defining an array.

```
int[] a1={1,2,3,4,5};
int a1[]=new int[]{1,2,3,4,5};
```

The sizes of arrays are decided by the numbers of values in the brace-enclosed list. Below are examples of array initializing.

【Example 2-17】**Initialize arrays**

```
public class For {
  public static void main(String[] args) {
    int a1[]=new int[]{1,2,3,4,5};
    for(int x:a1)
      System.out.print(x+" ");
  }
}
```

Output:

1 2 3 4 5

【Example 2-18】**String array**

```java
public class For {
  public static void main(String[] args) {
    String[] a1={"one","two","three"};
    String a2[]=new String[]{"four","six","seven"};
    for(String x:a1)
      System.out.print(x+" ");
    System.out.print('\n');
    for(String x:a2)
      System.out.print(x+" ");
  }
}
```

Output:

one two three
four six seven

【Example 2-19】

```java
public class ArrayInit {
  public static void main(String[] args) {
    Integer[] a={new Integer(1), new Integer(2), 3,}; //the last comma is selectable
    Integer[] b=new Integer[]{11, 22, 33,};    //autoboxing
    for(Integer x:a)
      System.out.print(x+" ");
    System.out.print('\n');
    for(Integer x:b)
      System.out.print(x+" ");
  }
}
```

Output:

1 2 3
11 22 33

【Example 2-20】

```java
public class Test {
  public static void main(String[] ss) {
    for(String s:ss)
      System.out.print(s+" ");
  }
}
```

2.9.3 Arrays act as arguments of methods（数组做方法的参数）

Array, class and interface variables are references, and can act as arguments of methods. Here are several examples of array arguments of methods.

【Example 2-21】Array arguments of methods

```java
public class Array {
  public static void main(String[] args) {
    Other ary=new Other();
    ary.method(new String[]{"Java", "is", "funny"});
  }
}
class Other {
  public void method(String[] ss) {
    for(String s:ss)
      System.out.print(s+" ");
  }
}
```

【Example 2-22】Array arguments of methods

```java
class A {}
public class VarArgs {
  void printArray(Object[] args) {
    for(Object obj:args)
      System.out.print(obj+" ");
    System.out.println();
  }
  public static void main(String[] args) {
    VarArgsarg=new VarArgs();
    arg.printArray(new Object[]{new Integer(47), new Float(3.14), new String("ok")});
    arg.printArray(new Object[]{"one", "two", "three"});
    arg.printArray(new Object[]{new A(), new A(), new A()});
  }
}
```

Where, the Object is the root class of all classes, so it can reference to all kinds of objects.

2.10 Command line arguments（命令行参数）

Parameters following commands are called command line parameters. The arguments of main method indicate that the actual parameters will come from the command line. While main is calling, the command line parameters will be passed over to main's arguments. String[] indicates the argument type is a String array.

跟在命令后面的参数是命令行参数。main 的实参来自命令行，调用 main，命令行参数被赋值给 main 的形参。String[] 指明形参是 String 数组。

【Example 2-23】Command line arguments

```java
public class HelloWorldApp {
  public static void main(String[] args) {
    System.out.println("Hello World!");
```

```
        System.out.println(args[0]);
        System.out.println(args[1]);
    }
}
```

Compile command:

javac HelloWorldApp.java

Running command：

Java HelloWorldApp abc

It will cause running errors, because the length of array args is the number of the command line parameters. The actual number of parameters, or array elements, is less than two, but the program takes two.

java HelloWorldApp cmd0 cmd1

or

java HelloWorldApp cmd0 cmd1 com2

Where cmd0 and cmd1 are passed to args[0] and args[1] respectively.

Output:

Hello World!

cmd0

cmd1

Dealing with command line parameters with foreach statement is suggested, because it won't take elements out of an array. Programs will run normally with or without command parameters.

建议用 foreach 语句处理命令行参数，因为 foreach 语句不会访问数组以外的元素，无论命令行有没有参数，有多少个参数，程序都能正常运行。

Exercises

1. Write a program to print default values of boolean, int and double fields. 输出默认值。
Hint:

(1) Define three fields. 定义 3 个成员变量。

(2) Define three methods, each returns a field. 定义 3 个方法，分别返回成员变量。

(3) Create an object (instance), call the three methods and print the calling results in main. 在 main 方法里创建一个对象（实例），分别调用 3 个方法并输出方法的返回结果。

2. Write a program to print default values of String, Byte fields and the other fields defined in the class. 输出 *String*，*Byte* 和这个类的其他成员变量的默认值。

3. Write a program to print the coercion results from int (value beyond range of byte) to byte. 写程序，输出从 *int*（值超过 *byte* 的范围）强制转换成 *byte* 的结果。

4. *Write a program to print the Unicode of a Java character.* 写程序，输出一个字符的 *Unicode* 码。

5. *Write a program.*

(1) In main method, define an int array, assign 1 to 100 to its items, use for statement. 在 main 里定义一个 int 型数组，用 for 语句将 1 到 100 赋值给数组元素。

(2) Output all items in the array, 10 items every line, use foreach and continue statements. 用 foreach 和 continue 语句按每行 10 个元素输出数组元素。

6. *Write a program.*

(1) Define a method named "fun" with an array type argument. In the method, increase 10% to each array element (use foreach loop). 定义一个叫 fun 的方法，形参是数组。方法将每个数组元素的值增加 10%（用 foreach）。

(2) In main, create an array of double, with initial value {27.6, 31.2, 28, 31.5, 31}. Print the array and invoke method fun, with the array as the actual parameter. Print the array again. 在 main 方法里创建一个 double 型数组，输出数组。调用方法 fun，用数组做实参，再次输出数组。

7. *Write a program (* 在 *cmd* 窗口执行 *).*

(1) Accept an argument from the command line.

(2) If the argument is "优" prints Excellent.

   If the argument is "良" prints Good.

   If the argument is "中" prints Average.

   If the argument is "不及格" prints Fail.

   If the argument is others prints Error.

Hint：

Using switch(args[0]).

8. *Write a program to print all command line parameters. (* 在 *cmd* 窗口执行 *)* 输出命令行参数。

Chapter 3
Classes and Objects
（类和对象）

3.1 Concepts of OOP（面向对象的概念）

Java supports object-oriented programming. This chapter studies the concept and structure of a class and how to create objects with the class.

3.1.1 Everything is an Object（万物皆对象）

The object-oriented programming takes everything as an object. The world is composed of objects. An object can be a computer, a person, or an intangible thing, such as a bank account. A program is a bunch of objects telling each other what to do by sending messages.

An object has state, behavior and identity. This means that an object can have internal data (state), methods (behavior), and each object can be uniquely distinguished from each other by its name, each object occupies a unique storage.

Abstraction is an important property of all object oriented programming languages. The same types of objects have the same properties, so we can extract the similar features from those objects. This approach is called abstraction. The concept of abstraction is to describe a set of similar concrete entities. For instance, as a whole, a car is a single object, in detail, the car consists of several subsystems: steering, brake, sound system, airconditioning and so on. In turn, each of these subsystems is made up of more specialized units.

面向对象的思想将一切事物都看成对象。大千世界由对象组成，对象可以是计算机、人或者银行账号这样无形的事物。程序包含若干对象，对象之间通过接口相互传递消息。

对象有内部数据（状态）、方法（行为），对象通过名字彼此区别，每个对象都在内存里占据唯一的存储空间。

抽象是面向对象程序设计语言的一个重要特性。同类对象有相似的特性，因此可以将相似的特性抽取出来，叫做抽象。抽象是对同类实体特征的描述。例如，小汽车是一个对象，小汽车包含多个子系统，驾驶系统、制动系统、音响系统、空调系统等，而每个子系统又由更为具体的单元组成。

3.1.2 Defining classes（定义类）

In chapters one and two, we've already seen some classes and used the "new" operator to create objects. Virtually all object-oriented programming languages use the "class" keyword.

```
class ClassName {
  … //fields
  … //constructors
  … //methods
}
```

This is a class definition. The class body contains declarations of the fields that provide the state of objects, and methods to implement the behavior of the objects. Fields represent member variables in a class. Constructors are methods too, for initializing new objects created from the class. Once a class is established, you can make as many objects of that class as you like.

3.2 Useful classes（常用类）

Java class library contains many classes. Here introduce several of them for use.

1. Data type corresponded classes

Each primitive data type corresponds to a class, for instance:

byte	Byte
int	Integer
char	Character
double	Double

…

All numerical type related classes contain parse methods to convert Strings to numerical data.

【Example 3-1】**Data type correspond classes**

```
public class TypeMaxValue {
  public static void main(String[] a) {
    byte maxByte=Byte.MAX_VALUE;
    short maxShort=Short.MAX_VALUE;
    int maxInt=Integer.MAX_VALUE;
    System.out.println("The maximum byte value: "+maxByte);
    System.out.println("The maximum short value: "+maxShort);
```

```
    int i=Integer.parseInt("123");
    double d=Double.parseDouble("12.5");
  }
}
```

2. Date class

Date class is available in java.util package. This class encapsulates the current date and time. The Date class provides two constructors:

(1) Date(), initializes an object with the current date and time.

(2) Date(long millisecond), accepts an argument that equals the number of milliseconds that have elapsed since midnight, January 1, 1970.

Java 的 java.util 包里有个 Date 类，该类封装了当前日期和时间。Date 类有两个构造方法：一个是 Date() 用当前时间和日期来初始化对象；另一个是 Date(long millisec) 含一个参数，其值代表从 1970 年 1 月 1 日午夜起流逝的毫秒数。

【Example 3-2】**Data class**

```
import java.util.Date;                    //or import java.util.*;
public class ClassDate {
  public static void main(String[] a) {
    Date date=new Date();
    System.out.println(date.toString()); //or System.out.println(date);
  }
}
```

Output:

Sun Sep 09 17:06:42 CST 2018

(CST 美国中央时区)

The current time and date is converted to a string of String type by toString() method. When the actual parameter of the println or print method is an object, the toString() can be omitted, that is,

```
System.out.println(date.toString());
```

Can be:

```
System.out.println(date);
```

E.g.

```
Test test=new Test();
    …
System.out.println(test.toString());
```

Can be:

```
System.out.println(test);
```

Except for the Date object, println or print method prints the address of an object if its actual parameter is an object. The address of an object consists of class name, @ and hash storage address of the object.

toString() 方法将当前日期和时间转换成 String 类型的字符串。当 println 或 print 方法的实参是对象时，可以省略 .toString()，即

```
System.out.println(date.toString());
```

可以写成：

```
System.out.println(date);
```

除了 Date 对象，如果 println 或 print 方法的实参是对象，它们输出对象的地址。对象的地址由"类名、@ 和对象的哈希存储地址"组成。

【Example 3-3】**Print an object of Date class**

```
import java.util.Date;
public class ClassDate {
  public static void main(String[] a) {
    System.out.println(new Date());
  }
}
```

3. String class

The String class has many methods, three of them are as follows:

```
(1) public char charAt(int arg0)
//Return the character at position arg0 of a string
(2) public String[] split(String arg0)
//Split a string into several String arrays according to arg0*/
(3) public char[] toCharArray()
//Put string elements to a char array
```

【Example 3-4】**String class**

```
class TryString {
  public static void main(String args[]) {
    String str=new String("hello!");
    char ch=str.charAt(0);                    //h
    System.out.println(ch);
    char[] sc=str.toCharArray();
    for(char s:sc)
      System.out.print(s);                    //hello!
    System.out.println();
    String[] string=str.split("l");
    for(String s:string)
      System.out.print(s+" ");                //he o!
    System.out.println();
    System.out.println(string.length);        //3
  }
}
```

Output:

h
hello!
he o!
3

Operator "+" and "+=" concatenates two strings or a string and any objects or primitive variables or any types of constants. The results are String type strings.

"+" 和 "+=" 运算符连接 2 个字符串，或者将一个字符串与任意对象、基本类型的任意变量、任意类型的常数连接，结果是 String 类型的字符串。

【Example 3-5】Operator "+"

```
class STR {
    public static void main(String args[]) {
        String str="hello!";
        str=str+" "+12+" "+67.2;
        str=str+" "+true;
        Ty ty=new Ty();
        str=str+" "+ty;  //str=str+ty.toString
        System.out.println(str);
    }
}
class Ty {}
```

Output:

hello! 12 67.2 true Ty@1fa1bb6

4. Keyboard input: Scanner class

Scanner class is available in java.util package too. Scanner class provides several methods to accept keyboard inputs of all primitive types except for char type. The char input can be realized by method next(), which accepts a string from the keyboard.

【Example 3-6】Scanner class

```
import java.util.Scanner; //import java.util.*;
public class Test {
  public static void main(String[] args) {
  Scanner input=new Scanner(System.in);
    System.out.println("请输入星期几");
    int today=input.nextInt();
    switch(today){
      case 1:
      case 3:
      case 5: System.out.println("English");
            break;
      case 2:
```

```
        case 4:
        case 6: System.out.println("Chinese");
                break;
        case 7: System.out.println("Japanese");
    }
  }
}
```

【Example 3-7】**Scanner class**

```
import java.util.Scanner;
public class ScannerTest {
  public static void main(String[] args) {
    Scanner input=new Scanner(System.in);
    System.out.println("请输入星期几");
    char today=input.next().charAt(0); //next() for String
    switch(today) {
      case '1':
      case '3':
      case '5': System.out.println("去工作");
                break;
      case '2':
      case '4':
      case '6': System.out.println("去上课");
                break;
      case'7': System.out.println("休息");
      default:System.out.println("必须是1-7");
    }
  }
}
```

【Example 3-8】**Accept an arbitrary number of integers from the keyboard and output them in reverse order. Where, an array is created dynamically.**

```
import java.util.Scanner;
public class Reverse {
  public static void main(String[] args) {
    int total;
    Scanner input=new Scanner(System.in);
    System.out.print("Enter the total number you'll input:");
    total=input.nextInt();
    int[] ary=new int[total];
    System.out.print("please input "+total+" integers:");
    for(int i=0;i<total;i++) {
      ary[i]=input.nextInt();
    }
    for(int i=total-1;i>=0;i--)
      System.out.print(ary[i]+",");
  }
}
```

3.3 Method overloading(方法重载)

Polymorphism is another property of OOP, which is accomplished by method overloading. Methods within the same class taking the same name but differing in arguments are termed overloaded methods.

Overloading means several methods take the same name. When method calling, the being called method is identified during the compile time according to its arguments. This is the static polymorphism.

多态，是 OOP 程序设计的另一个特点，通过方法重载实现。一个类里的多个方法可以取相同的名字，但方法的参数不同，叫做方法重载。

重载意味着多个方法取同一个名字。调用方法时，被调用的那个方法，是在编译时根据方法的参数确定的，是静态多态。

The overloaded methods share one name, but differ in arguments means the argument lists could differ at: number of arguments, data type of arguments, sequence of arguments.

重载的方法共用同一个名字，但它们的参数不同。参数不同指的是：参数个数，参数类型，参数顺序。

Have a look at the overloaded methods println, they are different in argument types, which is as shown in Figure 3-1. With println, almost all types of data can be outputted to the screen.

```
System.out.println
    println() : void - PrintStream
    println(boolean arg0) : void - PrintStream
    println(char arg0) : void - PrintStream
    println(char[] arg0) : void - PrintStream
    println(double arg0) : void - PrintStream
    println(float arg0) : void - PrintStream
    println(int arg0) : void - PrintStream
    println(long arg0) : void - PrintStream
    println(Object arg0) : void - PrintStream
    println(String arg0) : void - PrintStream
```

Figure 3-1 Overloaded methods println

For instance:
void add (int a, int b)
void add (int a, float b)

void sub (int a, float b)
void sub (int a, int b, float c)

void mult (int a, float b)

int mult (int a, float b)

The last pare of methods can't realize the method overloading by returning different method types.

【Example 3-9】**Method overloading**

```
class OverLoading { //static polymorphism
  void plan() {
    System.out.println("no arguments");
  }
  void plan(int x) {
    System.out.println("The value of x is "+x);
  }
}
public class OverLoad {
  public static void main(String args[]) {
    OverLoading ov=new OverLoading();
    ov.plan();
    ov.plan(5);
  }
}
```

It's very clear which plan() is called, that is the static polymorphism.

3.4 Constructors（构造方法）

Constructors play important roles in objects' initialization. A constructor is a method. Without constructors, an object can't be initialized properly. A constructor shares the same name as its class. It does not return a value, not even void. It may or may not have arguments. With method overloading, several constructors may be defined in a class.

If a class has constructors, Java automatically calls one of the constructors when an object is created.

构造方法用于对象的初始化。构造方法首先是方法，没有它，就无法恰当地初始化对象。构造方法名与类名相同，没有返回值，即使 void 也不能用。它既可以是有参方法，也可以是无参方法。方法重载，允许在一个类里定义多个构造方法。

如果类里有构造方法，创建对象时 Java 会自动调用其中的一个构造方法。

For instance:

```
public class Student {
  …
  Student() {…}
  Student(String Param) {…}
  …
}
```

【Example 3-10】 **Objects personl and person2 hold default values to their fields**

```
class Account {
  private int act_no;
  private double balance;
  private String name;
  public void display() {
    System.out.println("Account:"+act_no);
    System.out.println("Name:"+name);
    System.out.println("Name:"+balance);
  }
  public void deposit(double amount) {
    …
  }
  public void withdraw(double amount) {
    …
  }
  public static void main( String args[]) {
    Account person1=new Account();
    Account person2=new Account();
    Person1.display();
    Person2.display();
  }
}
```

Objects person1 and person2 hold default values to their fields. Different objects, but hold the same field values. Using constructors, a newly created object can hold individual initial values. If a class has constructors, Java automatically calls one of the constructors when an object is created.

对象 person1 和 person2 的数据都具有默认值，不同的对象却有相同的值。使用构造方法，新建对象的 fields 可以有自己特定的初始值。如果类里有多个构造方法，创建对象时，Java 会自动调用其中的一个。

Let's redo example 3-10. By defining a constructor, each person can have his/her own individual account.

【Example 3-11】 **Assignment and output an object**

```
import java.util.Scanner;
public class Account {
  private int act_no;
  private double balance;
  private String name;
  public void display() {
    System.out.println("Account:"+act_no);
    System.out.println("Name:"+name);
    System.out.println("Balance:"+balance);
  }
  Account(int acc, String nm, double b) {
    act_no=acc;
```

```
      name=nm;
      balance=b;
    }
    public static void main(String[] args) {
      int act;
      String name;
      double b;
      Scanner in=new Scanner(System.in);
      System.out.println("Account:");
      act=in.nextInt();
      System.out.println("Name:");
      name=in.next();
      System.out.println("Balance:");
      b=in.nextDouble();
      Account person=new Account(act, name, b);
      person.display();
    }
}
```

Input:

Account: 12

Name: Jenney

Balance: 120.5

Output:

Account: 12

Name: Jenney

Balance: 120.5

【Example 3-12】**Constructor overloading**

```
class Tree {
  int height;
  public Tree() {          //constructor
    System.out.println("Planting a seed");
    height=0;
  }
  public Tree(int i) {     //overloading constructor
    System.out.println("Creating new Tree " + i + " feet tall");
    height=i;
  }
}
```

【Example 3-13】**Assignment and output an object**

```
class DataOnly {
  int i;
  float f;
  boolean b;
```

```
  public DataOnly( int x, float y, boolean z ) { i=x;f=y;b=z; }
  public static void main(String[] args) {
    DataOnly d=new DataOnly(1, 2, true);
    System.out.println("i="+d.i);
    System.out.println("f="+d.f);
    System.out.println("b="+d.b);
  }
}
```

Output:

i= 1

f= 2.0

b= true

【Example 3-14】**Constrctor method**

```
class Rock {
  public Rock(int k) {
    System.out.println("Create rock"+k);
  }
}
public class Constructor {
  public static void main(String[] args) {
    for(int i=0;i<5;i++)
      new Rock(i);
  }
}
```

Output:

Create rock 0

Create rock 1

Create rock 2

Create rock 3

Create rock 4

3.5 Default constructor（默认构造方法）

A constructor that takes no arguments is called default constructor. But like any methods, a constructor can also take arguments.

无参构造方法叫默认构造方法，与其他方法一样，构造方法也可以有参数。

【Example 3-15】**Default constructor**

```
class Rock {
  int k=100; //same value for all objects
  public Rock() {
```

```
    System.out.println("Creating Rock k="+k);
  }
public class Constructor {
  public static void main(String[] args) {
    for(int i=0;i<5;i++)
    new Rock();
  }
}
}
```

Output:

Creating Rock k=100
Creating Rock k=100
Creating Rock k=100
Creating Rock k=100
Creating Rock k=100

The default constructor Rock() is called five times. If a class has no constructors, a default constructor will be automatically provided and be called while creating an object.

构造方法 Rock() 被调用了 5 次。如果未在类中定义构造方法，Java 会自动创建一个默认构造方法，并在创建对象时自动调用。

【Example 3-16】**Default constructor**

```
class Bird {
  //no constructor
  //…
}
public class DefaultConstructor {
  public static void main(String[] args) {
    Bird nc=new Bird(); //invoke default constructor Bird(){}
    DefaultConstructor dc=new DefaultConstructor(); //invoke
    //DefaultConstructor(){}
  }
}
```

Two default constructors Bird(){} and DefaultConstructor(){} are provided respectively, and one of them will be invoked while a "new" statement is executed. However, if any constructors (with or without arguments) are defined in a class, the compiler will not provide any for you.

系统为两个类分别提供了默认构造方法 Bird(){} 和 DefaultConstructor(){}，在执行 new 语句时，自动调用。如果已经在类中定义了构造方法（有参或者无参），编译器将不再提供默认构造方法。

【Example 3-17】**Constructor overloading**

```
class Bird {
  Bird(String color) {…}
```

```
    Bird(char color, String name) {…}
}
…
Bird a=new Bird(); //error
Bird b=new Bird("Hello");
Bird c=new Bird('s', "Hello");
```

3.6 Static fields and methods(静态成员变量与静态方法)

Method main is a static method. In the static methods, non-static fields and methods of the class can't be referenced directly, you must create an object (instance) and use that object to access to the fields or methods, since non-static fields and methods must know the particular object they are working with. When you define a class, you don't actually get an object until you create one using the "new" keyword, and at that point storage is allocated for the fields and the methods are callable. In example 3-18, the non-static method "noStatic" is invoked in main method by the object "ns" of Student and the non-static field "anon" is assigned a value by the object "ns" too, which indicates that the "anon" is the field that stays at the storage of "ns", but not the storage of "ms", and the method "noStatic" prints the "anon" of "ns", but not the one of "ms", and the "anon" of "ms" keeps its default value unchanged.

main 是静态方法，在静态方法中不能直接访问类的非静态的成员变量和方法，必须先创建对象，并通过指定的对象访问非静态的成员变量和方法。定义类，并通过 new 关键字创建对象后，成员变量才有存储空间，方法才可以被调用。在例 3-18 的 main 方法里，非静态方法"noStatic"是通过 Student 的对象"ns"调用的，非静态成员变量"anon"也是通过"ns"赋值的，这说明"anon"是"ns"的，而不是"ms"的。方法"noStatic"使用的数据是"ns"的"anon"，而不是"ms"的"anon"，"ms"的"anon"还是原来的默认值。

【Example 3-18】Non-static fields and methods

```
class Student {
  int anon;                               //non-static field
  void noStatic() {                       //non static method
    System.out.println(anon);
  }
  public static void main(String c[])    { //static method
    Student ns=new Student();
    Student ms=new Student();
    ns.anon=12;
    ns.noStatic(); //create an object first, then access to member method
    System.out.println(ms.anon);
  }
}
```

Output:

12

0

In some cases, calling a method without creating an object is necessary. The main, entry point for running an application, is called without any object. You may also want to have only a single piece of storage for a particular field, regardless of how many objects of that class are created, or even if no objects are created. A field, like total number of employees, needs only one copy, unlike employee's name, each object should have its own name. You may need a method that isn't associated with any particular object of its class, that is, you need a method that you can call even if no objects are created. You can achieve both of these effects with the "static" keyword. When you say a field or a method is static, it means a particular field or a method is not tied to any particular object of that class. So even if you've never created an object of that class, you can call a static method or access to a static field. To make a field or a method static, you simply place the keyword "static" before a definition. In example 3-19, the static method "isStatic" is invoked in main method directly and the static field "isst" is assigned a value directly too, which indicate that the "isst" stands alone and relates none of objects, and the method "isStatic" prints the only one "isst", it is there without creating any objects.

有时调用方法，不需要创建对象。例如，在调用应用程序的入口 main 方法时并没有创建任何对象。你也可能希望某个类的成员变量不属于任何对象，即只有一个，与创建了多少对象，甚至是否创建对象无关。像员工总数这样的成员变量，只需要一个数据，而不像员工的姓名那样，每个对象都有一个员工的名字。你可能希望不通过对象就调用方法。使用 static 关键字，上述两者都能实现。说某成员变量或某方法是静态的，意味它们与特定的对象无关，即使未创建对象也可以使用它们。若让方法或成员变量成为静态的，只需在它们前面加 static。例 3-19 中，main 方法直接调用静态方法 "isStatic"，成员变量 "isst" 也直接被赋值。这说明 "isst" 独立存在，与任何对象无关。方法 "isStatic" 输出唯一的 "isst"，无须创建任何对象。

【Example 3-19】**Static fields and methods**

```
class Student {
  static int isst=72;                   //static field
  static void isStatic() {              //static method
    System.out.println(isst);
  }
  public static void main(String c[]) { //main is static
    isStatic(); //access method without any object
    isst=12;
    System.out.println(isst);
  }
}
```

Output:

72

12

We call static fields "class fields", and call static methods "class methods", meaning that the fields and methods exist for the class, and not for any particular objects of the class. On the other hand, call non-static fields "instance fields", and non-static methods "instance methods". Actually, an object is an instance. Some developing tools even prefer instance to object. All fields and methods (static or non-static) can be dealt with and invoked directly within a non-static method. Later on we'll tell how the non-static fields and methods in a non-static method relate to an object. In example 3-20, the static method "fun" is called both in the static method main and the non-static method nsFun, however the non-static members are different, within the static method main, they must relate to objects explicitly, but in the non-static method nsFun, they do relate to an object, but in a hidden way.

把静态成员变量和静态方法分别称为类变量和类方法,即它们属于整个类而不是特定的对象。而把非静态成员变量称为实例变量,非静态方法称为实例方法。事实上,对象与实例的含义相同,有些开发工具更喜欢用实例而不用对象。非静态方法能够直接访问所有成员变量,直接调用所有方法(静态或非静态的)。以后,我们会介绍非静态方法里的非静态的成员变量和方法是如何与对象关联的。在例3-20里,静态方法main和非静态方法nsFun都调用了静态方法fun,可是对于非静态成员就不同了,在静态方法main里它们必须与对象关联,而在非静态方法nsFun里,它们的确与某个对象有关,但却是以一种隐含的方式关联的。

【Example 3-20】**Static fields and methods**

```
public class StaticTest {
  int ak, bk; //non-static
  static int stc=47; //static
  public static void main(String c[]) { //static
    fun(); //access to static method without any object
    StaticTest st1=new StaticTest();
    StaticTest st2=new StaticTest();
    st2.ak=12;
    st2.bk=19;
    System.out.println("Object st1: "+st1.ak+" "+st1.bk);
    //access to non-static fields with object
    System.out.println("Object st2: "+st2.ak+" "+st2.bk);
    st2.nsFun();
    System.out.println("Static: "+StaticTest.stc+" "+stc);
    //access to a static field without object
  }
  public static void fun() {
    System.out.println("Static in fun: "+stc);
  }
  public void nsFun() {
```

```
      fun();  //call static method
      System.out.println("Static in nsFun: "+stc);   //static field
      System.out.println("Non-static in nsFun: "+ak); //non-static field
   }
}
```

Output:

Static in fun: 47

Object st1: 0 0

Object st2: 12 19

Static in fun: 47

Static in nsFun: 47

Non-static in nsFun: 12

Static: 47 47

Only one storage for static field stc, and each object has its own storage for its fields ak and bk respectively (See Figure 3-2).

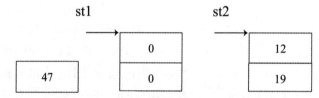

Figure 3-2 One storage for static field

There are three approaches to refer to a static member of a class (see example 3-21):

(1) A reference that refers to an object.

(2) Directly by name of the static fields or methods.

(3) The class name, where the static fields or methods are defined.

If static fields and methods are defined in a class are used in other classes, the name of the class where they are defined should be used.

有3种访问静态成员的方式（见例3-21）：

（1）指向对象的引用；

（2）直接使用静态成员的名字方法；

（3）静态成员所在类的类名。

当静态成员在另一个类中被访问时，则需用定义静态成员的那个类的类名访问。

【Example 3-21】Three approaches to refer to a static member of a class

```
class X {
  public static void main(String[] k) {
    Y a=new Y(); Y b=new Y(); Y c=new Y(); Y d=new Y();
    System.out.println("x="+a.x+" "+b.x+" "+c.x+" "+d.x+" "+Y.x);
    a.prtX();                                 //by reference
```

```
      Y.prtX();                        //by class name
   }
}
class Y {
   static int x=12;
   static void prtX() {
      System.out.println("x:"+x);       //by name
   }
}
```

Output:

x=12 12 12 12 12

x:12

x:12

You've seen that no any difference for the static member is referenced to by which object a, b, c or d, they are just the same as referenced by the class name Y. Similar logic applies to static methods. The x belongs to class Y, not any objects of class Y.

【Example 3-22】**Static fields and methods**

```
public class StaticTest {
   static int i=27;
   public static void main(String[] args) { //static
      StaticTest obj=new StaticTest();
      obj.function();     //call non-static method, by object
      method();           //call static method directly by name
      AnotherClass another=new AnotherClass();
      another.nonstatic();//call non-static method in different class by object
      another.increment(); //call static method in different class by object
      AnotherClass.increment(); //or by name of that class
   }
   static void method() {
      System.out.println("I'm a static method");
   }
   static void function() {
      StaticTest st1=new StaticTest();
      StaticTest st2=new StaticTest();
      st1.i=12; //reference to static i by object
      System.out.println("i="+i); //reference to static i directly by name
      st2.i=38; //reference to static i by different object
      System.out.println("i="+i);
      StaticTest.i++; //reference to static i by class name
      i++; //reference to static i directly by name
      System.out.println("i="+i);
   }
}
class AnotherClass {
   void nonstatic() {
      StaticTest.i++;
```

```
        increment();  //call static method in this class directly by name
        System.out.println("i="+StaticTest.i);
        //class name, static i in another class
        StaticTest.method(); //by class name, static method is in another class
        method();            //error
    }
    static void increment() {
        StaticTest.i++;
    }
}
```

Output:

i=12

i=38

i=40

I'm a static method

i=42

I'm a static method

【Example 3-23】**Static fields and methods**

```
class Cup {
  int a,b;
  Cup(int x, int y) {a=x; b=y;} //use constructor
}
class Cups {
  static Cup cup1=new Cup(1,2);
  static Cup cup2=new Cup(11,22);
  /*
  static Cup cup1;
  static Cup cup2;
  static { //static block
    cup1=new Cup(1,2);
    cup2=new Cup(11,22);
  }
  */
}
public class Init {
  public static void main(String[] args) {
    System.out.println(Cups.cup1.a);
    System.out.println(Cups.cup1.b);
    System.out.println(Cups.cup2.a);
    System.out.println(Cups.cup2.b);
  }
}
```

Output:

1

2

11
22

3.7　This keyword（this 关键字）

All members in a class can be reference to directly in a non-static method, which is realized by keyword "this". The "this" is a hidden argument of a non-static method and is a reference that will refer to an object when it is assigned with an object.

类的非静态方法能够直接访问类里的所有成员，这是通过 this 实现的，this 是非静态方法的一个隐藏参数，是引用，被赋值后将代表具体的对象。

3.7.1　A non-static method has a hidden "this"（隐藏参数 this 的非静态方法）

【Example 3-24】**Keyword this**

```
public class Student {
  private int stuID;
  public Student(int stuID){
    this.stuID=stuID;
  }
}
```

In example 3-24, the field and the constructor's argument take the same name stuID. Without "this", the two stuIDs in the constructor would be confused. The "this" is a hidden argument of a non-static method. Its data type is the present class, here, Student.

例 3-24 中，field 和构造方法的参数都是 stuID。如果没有 this，构造方法里的两个 stuID 就混淆了。this 是非静态方法的一个隐藏参数，其类型是当前类，这里是 Student。

【Example 3-25】**Keyword this**

```
public class Student {
  private int stuID;
  public Student (int stuID) {
  //equivalent to public Student(intstuID, Student this)
    this.stuID=stuID;
  }
  public int getStuID() {  //getStuID(Student this)
    return this.stuID;     //return stuID;
  }
  public void display() {
    System.out.println(stuID); //System.out.println(this.stuID);
  }
  public static void main(String[] s) {
    Student ak=new Student(101);
  //ak is passed to the hidden"this"of the constructor
    Student bk=new Student(202); //the hidden"this" has the same value as bk
```

```
            ak.display();
            bk.display();
            System.out.println(ak.getStuID());
        }
    }
```

Output:

101

202

101

Different objects (actual parameters) pass different objects to "this". When the statement Student ak=new Student(101); is executed, the constructor is called, and ak (storage address of the new object) is passed to "this" of the constructor, and the value of "this" equals ak. When statement Student bk=new Student(202); is executed, "this" equals bk (See Figure 3-3).

不同对象（实参）传递不同的对象给 this。执行语句 Student ak=new Student(101); 时，构造方法被调用，ak（新对象的存储地址）传递给构造方法的 this。同样地，执行语句 Student bk=new Student(202); 时，bk 传递给 this。

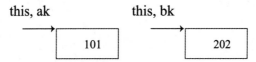

Figure 3-3 "this" of the constructor Student is assigned

【Example 3-26】**Object assignment**

```
class Foo {
    private boolean x, y=true;
    boolean getOR(boolean a, boolean b) {
        x=a; y=b;                //this.x=a; this.y=b;
        System.out.println("this equals object="+this);
        return x||y;
    }
}
public class TestThis {
    public static void main(String[] a) {
        Foo foo1=new Foo();    //this=foo1
        Foo foo2=new Foo();    //this=foo2
        System.out.println("this equals foo1="+foo1);
        System.out.println("x OR y="+foo1.getOR(false, false));
        System.out.println("this equals foo2="+foo2);
        System.out.println("x OR y="+foo2.getOR(true, false));
    }
}
```

The "this" is a hidden method argument. It does not appear on the argument list of a method. When a non-static method is invoked, the storage address of the object is automatically assigned to "this".

"this"是隐藏参数。它并不出现在方法的参数列表中。调用方法时，当前对象的存储地址被自动地赋值给 this。

【Example 3-27】**Object assignment**

```
class Banana {
  int numb;
  void peel(int i) {
    numb=i;
    System.out.println(numb);
  }
}
public class BananaPeel {
  public static void main(String[] args) {
    Banana ak=new Banana();
    Banana bk=new Banana();
    ak.peel(11); //this=ak, this.numb=11 means ak.numb=11
    bk.peel(22);
  }
}
```

When the statement ak.peel(11); is executed, the value of ak is passed to the hidde "this" of peel, and "numb" of ak is assigned 11. Calling method peel, without "this", how can that peel know who's "numb"? object ak or bk should be assigned?

当执行 ak.peel(11);语句时，ak 的值被赋值给 peel 的隐藏 this，因此，ak 的 numb 被赋值 11。如果没有 this，调用 peel 方法时，peel 就不知道是给对象 ak 的 numb 赋值还是给 bk 的 numb 赋值。

【Example 3-28】**Serially call a method using "this" keyword**

```
public class Leaf {
  int i=0;
  Leaf increment() {
    i++; //this.i++
    return this;
  }
  void print() {
    System.out.println("i="+i);
  }
  public static void main(String[] args) {
    Leaf x=new Leaf();
    x.increment().increment().increment().print(); //x.print();
  }
}
```

Output:

i = 3

3.7.2　A static method has no argument "this"（没有 this 的静态方法）

A static method has no argument "this", so in a static method if you never created an object of that class you can only invoke static methods and access to static fields.

Without "this", static methods can't reference to any objects, can only directly access to static members (static methods, static fields), but can't directly access to non-static members (methods, fields) of a class.

静态方法没有 this 参数，所以如果未在静态方法里创建对象，用户只能访问静态方法和静态数据。

由于没有 this，静态方法不能引用任何对象，只能直接访问静态成员（静态方法、静态成员变量），而不能直接访问类的非静态成员（方法、成员变量）。

【Example 3-29】**Static method main can only directly access to static members**

```
class Test {
  static private int a;
  private int b;
  public void print() {
    System.out.println("a="+a);
    System.out.println("b="+b);
  }
  public static void main(String args[]) {
    System.out.println("a="+a);
    System.out.println("b="+b); //error
  }
}
```

【Example 3-30】**Static and non-static members**

```
class TestStatic {
  static int ak;
  private int bk;
  public static void print() {
    System.out.println(" ak="+ak);
    System.out.println(" bk="+bk); //error
  }
}
```

For lack of "this", static method print() can't tell bk, non-static field, belongs to which object! Making correction as example 3-31, bk can be accessed indirectly.

由于没有 this，静态方法 print() 无法指出非静态成员 bk 属于哪个对象。例 3-31 中做如下改动，print() 可以间接地访问 bk。

【Example 3-31】**Static method**

```
public static void print () {
  System.out.println("ak="+ak);
```

```
    TestStatic y=new TestStatic();
    System.out.println("bk="+y.bk);
}
```

【Example 3-32】**Static method main indirectly accesses to non-static members of class Add**

```
class Add {
  int num1, num2;
  public int addNumber() {
    int sum;
    sum=num1+num2;
    return sum;
  }
}
public class ClassDemo {
    public static void main(String args[]) {
    Add obj=new Add();
    obj.num1=1;
    obj.num2=2;
    System.out.println("The sum is: "+obj.addNumber());
    }
}
```

【Example 3-33】**Static and non-static methods**

```
class A {
  public void q() {
    System.out.println("I'm A!");
    B.p();
  }
}
class B {
  public static void p() {
    System.out.println("I'm B!");
  }
  public void r() {
    p();
  }
}
public class TestStatic {
    public static void main(String[] a) {
    B.p(); //without instantiation
    A k=new A();
    k.q(); //must be instantiated
    B kk=new B();
    kk.r();
    }
}
```

Output:

I'm B!
I'm A!
I'm B!
I'm B!

3.7.3 Calling constructors form constructors（在构造方法里调用其他构造方法）

You may define several constructors in a class, when you'd like to call one constructor from another to avoid duplicating codes. You can do it using keyword "this".

你可能定义了好几个构造方法，为了避免代码重复，希望通过另一个构造方法来调用其他的构造方法，用 this 关键字可以实现这个想法。

【Example 3-34】 **Calling constructors form constructors**

```
class Flower {
  private int count=0;
  private String s=" 花瓣 ";     //String s=new String(" 花瓣 ");
  public static void main(String args[]) {
    Flower b=new Flower();
    System.out.println(b.s+' '+b.Count+' 个 ');
  }
  Flower(int petals) {
    count=petals;
  }
  Flower() {
    this(10);
  }                                //avoid duplicating code
}
```

3.8 Variable argument lists（可变参数列表）

A variable argument list of a method can include unknown quantities of arguments.
Syntax: type … variable
方法的参数列表里的参数个数是可变的。
int fun1(int … x) {}
String fun2(Object … y) {}
Object fun3(Integer … z) {}

【Example 3-35】 **Variable arguments**

```
public class Arguments {
  static void fun(int k, String ... kk) {
    System.out.print("k="+k+"\n");
```

```
      for(String s:kk)
        System.out.print(s+" ");
      System.out.println();
    }
    public static void main(String[] args) {
      fun(1, "one");
      fun(2, "two", "three");
      fun(0);
    }
}
```

Output:

k=1

one

k=2

two three

k=0

【Example 3-36】**Variable arguments**

```
class A {}
public class NewVarArgs {
  static void printArray(Object ... args) {
    for(Object obj:args)
    System.out.print(obj+" ");
    System.out.println();
  }
  public static void main(String[] args) {
    printArray(new Integer(1), new Float(1.1),"okay"); //individual elements
    printArray(11, 2.2);
    printArray("one", "two", "three", "four");
    printArray(new A(), new A(), new A());
    printArray((Object[])new Integer[]{1,2,3,4});
    //coerce Integer array to Object array
    printAr ray(); //empty list is OK
  }
}
```

Output:

1 1.1 okay

11 2.2

one two three four

A@19360e2 A@bdb503 A@b6e39f

1 2 3 4

In fact, method main can be rewritten with a variable argument list.

```
public class TestMain {
```

```
  public static void main(String ... args) {
    for(String k:args)
      System.out.print(k+" ");
  }
}
```

【Example 3-37】Variable argument and serial call

```
public class AddTo {
  static int k=0; //length of String array
  String[] ss=new String[50];
  public AddTo add(String ... vs) {
    for(String v:vs)
      ss[k++]=v; //assignment values to array
    return this;
  }
  public static void main(String[] args) {
    AddTo to=new AddTo();
    to.add("a","b");
    to.add("c","d","e").add("f").add("g","h","i","j"); //serial call
    int i=0;
    while(i<k) System.out.print(to.ss[i++]+" ");
  }
}
```

Output:

a b c d e f g h i j

3.9 Garbage collection（垃圾回收）

In C++ language, it is the programmer's responsibility to de-allocate the memory allocated dynamically, keywords "new" and "delete" are used to allocate and reclaim memory. Programmers know about the importance of initialization, but often forget the importance of clearing up and leaving some memory unreclaimed.

Java provides a garbage collector to reclaim the memory of objects that are no longer used. The garbage collector knows how to release memory allocated with "new". So you never need to destroy an object! The memory de-allocation is automatically done. This mechanism is called "Garbage collection".

Setting a reference to null means that is no longer referring to any object. An object which is not referred by any reference will be removed from memory by the garbage collector. All objects are living on the heap. Figure 3-4 illustrates the de-allocation procedure.

C++ 语言中，回收动态分配的存储空间的任务是通过程序员编写代码来实现的。关键字 new 和 delete 用于动态分配与回收存储空间。程序员都知道初始化的重要性，但常常忘记存储器回收的重要性，使得一些存储空间无法回收。

Java 提供垃圾回收程序，回收不再使用的对象的存储空间。垃圾回收程序知道如何释放通过 new 分配的存储空间。因此，Java 从不需要撤销对象！存储空间的回收是自动完成的，这种机制称为"垃圾回收"。

将 null 赋值给引用，说明该引用与任何对象无关。垃圾回收机制回收与任何引用无关联的对象。所有对象都在堆（heap）中，图 3-4 说明垃圾回收过程。

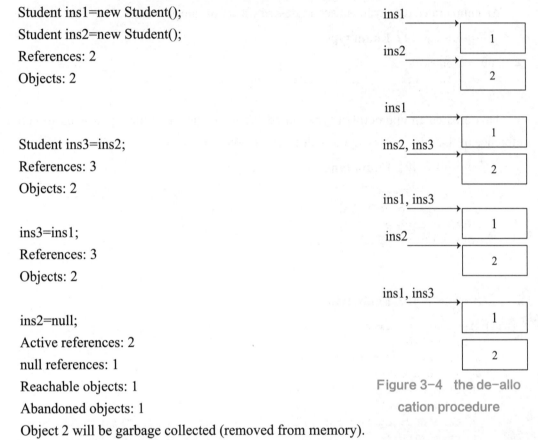

Student ins1=new Student();
Student ins2=new Student();
References: 2
Objects: 2

Student ins3=ins2;
References: 3
Objects: 2

ins3=ins1;
References: 3
Objects: 2

ins2=null;
Active references: 2
null references: 1
Reachable objects: 1
Abandoned objects: 1

Figure 3-4　the de-allocation procedure

Object 2 will be garbage collected (removed from memory).

【Example 3-38】**java.util.Date()**

```
public class HelloDate {
  public static void main(String[] a) {
    System.out.print("Hello, it's: ");
    System.out.println(new java.util.Date());
  }
}
```

Output:

Hello, it's: Tue Jul 10 09:26:52 CST 2018

Where, the actual argument is a Date object that is created just to send its value, which is automatically converted to a String to print. As soon as the statement is finished, that Date object is unnecessary, and the garbage collector can come along and get it anytime.

println() 方法的实参是 Date 类的对象，创建这个对象仅仅是为了获取其值（对象被自动转换成 String）。当这条语句结束，Date 对象将被垃圾回收。

3.10 Enum type（枚举类型）

An enumerated data type is used to describe a set of integer values.

【Example 3-39】**Enum type**

```
public enum Week {
  Mon, Tue, Wed, Thur, Fri, Sat, Sun
}
```

This creates an enumerated type called Week. To use an enum, you need to create a reference of that type and assign it with an enum constant.

【Example 3-40】**Enum type**

```
public class UseEnum {
  public static void main(String[] args) {
    Week today=Week.Tue;
    System.out.println(today);
  }
}
```

【Example 3-41】**Enum type**

```
public class Burrito {
  Spici degree;
  public Burrito(Spici d) {degree=d;}
  public void describe() {
    System.out.print("This burrito is ");
    switch(degree) {
      case NOT: System.out.println("not spicy at all."); break;
      case MILD:
      case MEDIUM: System.out.println("a little hot."); break;
      case HOT:
      case FLAMING:
      default: System.out.println("maybe too hot.");
    }
  }
  public static void main(String[] args) {
    Burrito plain=new Burrito(Spici.NOT), //逗号
    greenChile=new Burrito(Spici.MEDIUM),
    jalapeno=new Burrito(Spici.HOT);
    plain.describe();
    greenChile.describe();
    jalapeno.describe();
  }
}
```

Exercises

1. Write a program to accept several decimal numbers from the keyboard, output them in reverse order and output the average of them.

(1) Promote: "Enter the total number you'll input:"

(2) Accept inputs from the keyboard.

(3) Print in reverse order and the average of them.

写程序，从键盘接收几个小数，按反序输出，并输出它们的平均值。

(1) 提示："请输入小数的个数："。

(2) 接收数据。

(3) 反序输出，并输出平均值。

2. Write a program to accept characters one by one from the keyboard. If a character equals "N" or "n", output "I won't leave." Or else output "Must be Y or N", until a character equals "Y" or "y", output "Good Bye" and end the program.

写程序，从键盘输入字符，如果字符为"N"或"n"，输出"I won't leave."否则输出"Must be Y or N"，直到输入的字符为"Y"或"y"，输出"Good Bye"，并结束程序。

3. Write a program to create a class with 2 String fields. One is initialized with a constant string, and another is initialized by a constructor that takes an argument. Create an object in main to output the two strings. The actual parameter comes from the keyboard. Create another object, and output the two strings too. Pay attention to the difference between the two initializing approaches.

定义一个类，类中定义 2 个字符成员变量。其中一个用字符串常量赋初值，另一个通过一个含参数的构造方法赋初值。在 main 里创建一个对象，输出这 2 个成员变量，实参来自键盘。再创建一个对象，输出这两个成员变量。注意两种初始化方法的区别。

4. Write a program to define two classes, Teacher and TeacherTest, TeacherTest is the main class.

1) Class Teacher has two fields: name and age

(1) Define two constructors, one is default, assign constants to name and age fields respectively. Another constructor takes two arguments for initializing the two fields.

(2) Define method "introduction" as below.

```
public void introduction() {
System.out.println("Hello every body, my name is"+name+", my age is "+age+" years old.");
}
```

2) Class TeacherTest

(1) In main of Teacher Test, create two objects of Teacher, one will call the non-argument constructor, another will call the constructor taking arguments.

(2) Call method "introduction" through the two objects.

写程序，定义 2 个类，Teacher 和 TeacherTest，TeacherTest 是主类。

1) Teacher 有 2 个 fields：name 和 age

(1) 定义 2 个构造方法，其中一个是无参的，分别用 2 个常数给 name 和 age 赋值。另一个构造方法含 2 个参数，用于给 2 个 fields 赋值。

(2) 定义如下的 introduction 方法

```
public void introduction() {
  System.out.println("Hello every body, my name is "+name+", my age is "+age+ " years old.");
}
```

2) TeacherTest 类

(1) 在它的主方法里创建 2 个 Teacher 类的对象，一个调用缺省构造方法，另一个调用有参构造方法。

(2) 通过这 2 个对象调用 introduction 方法。

5. *Output the default of the static field in four different approaches.*（用四种不同的方法输出静态成员的默认值。）

```
public class Static {
  static String st;
  public static void main(String[] args) {
    Static s1=new Static();
    Static s2=new Static();
    …
    …
    …
    }
}
```

6. *Write a program.*

(1) Define a Student class, which contains fields: sId, sName, sTotalNo, and a constructor to initialize objects, the actual parameters comes from the keyboard.

(2) In main, create two objects and print the two students' information and the total number of students.

写程序：

(1) 定义 Student 类，类里包含 fields：sId, sName, sTotalNo 和一个构造方法。构造方法用来初始化对象，实参由键盘输入。

(2) 在 main 里创建 2 个对象，输出 2 个学生的信息和学生总数。

7. *Write a program to output an object. Call a method by the object. The method output "this" to see if the two outputs are equal.*

写程序，输出一个对象，通过这个对象调用方法，方法里输出 this，看 2 个输出是否相同。

8. Write a program to compute factorial of 5 by serially calling a method 4 times.
Hint:
fact(2). fact(3). fact(4). fact(5).getv();
计算 5 的阶乘，通过串行调用 1 个方法 4 次。
getv 方法输出阶乘的结果，类中只有一个变量，串行调用 fact 方法都是针对这个变量。
5!=1*2*3*4*5=120

9. Write a program.
(1) Define a method that takes int variable argument list and print all its parameters.
(2) Call the method 3 times, each time with different number of integers in the main method.
写程序：
(1) 定义一个具有 int 型可变参数列表的方法，输出所有参数。
(2) 在 main 里调用该方法 3 次，每次调用的整数个数都不同。

Chapter 4
Packages（包）

4.1 Concept of packages（包的概念）

A package corresponds to a directory in the file management system. Java uses a hierarchical file system to manage source and bytecode files too. A package is a collection of related classes (for instance, base classes and there subclasses), interfaces and/or other packages. Each package has its own name. Classes and interfaces with the same names cannot appear in the same package, they can appear in different packages.

包与文件管理系统中的文件目录对应，Java 也采用分层的文件系统来管理源文件和字节码文件。每个包都有自己的名字，包是相关类（例如，基类与其子类）、接口和（或）其他包的集合。同名的类和接口不能出现在同一个包里，但可以在不同的包中。

Packages allow you to organize related classes and interfaces into smaller units and make them easy to locate and use. The package mechanism is the naming space management, can avoid naming conflicts. Package names can be used to identify classes, that is, a full class name includes a package name, such as java.util.Date. By convention, companies use their reserved Internet domain name as their package names, such as: com.company.package.

包允许用户将类和接口组织成较小的单位，方便使用。包使用命名空间管理机制，能够防止名字冲突。包名可以用来标识类名，事实上，类名的全称包含包名，例如，java.util.Date。按惯例，一些公司用互联网的域名作为包名。例如，com.company.package。

4.2 Java library and its package structure（类库与 Java 类的包组织结构）

Java JRE provides a powerful class library for use. All built-in classes in the library are organized into several packages according to their functionalities.

At …\Java\jdk1.8.0\jre\lib\, you'll see a compressed jar file as shown in Figure 4-1. The rt.jar is an executable jar file and you can find all Java classes in it. Under the package (folder) "java", all subpackages (subfolders) are as shown in Figure 4-2. Some subpackages have been used and are going to be used in later chapters. Opening a subpackage, you'll see many bytecode (.class) files in it.

Figure 4–1 File rt.jar

JRE 提供了强大的类库。Java 以包的形式组织类库里的类，将类按用途分放到不同的包和子包里。

在…\Java\jdk1.8.0\jre\lib\ 下有个压缩文件 rt.jar，它是个可执行的 jar 文件，你能够在 rt.jar 文件里找到 Java 类库里的所有类。在它的 java 包（文件夹）里可以看到如图 4-2 所示的全部子包（子文件夹）。有的子包前面已经使用了，有的会在后面的章节里用到。打开子包，可以看到子包里的字节码（.class）文件。

Figure 4-2 Sub packages under package "java"

(1) java.lang includes language related classes, which is imported into each source program automatically.

(2) java.io includes input/output related classes, such as file dealing with classes.

(3) java.util includes useful classes, such as Date, Scanner classes.
(4) java.net includes network communication related classes.
…

（1）java.lang 包含与程序设计语言相关的类，该包被自动导入每个 Java 源程序。
（2）java.io 包含与输入 / 输出操作相关的类，如与文件相关的类。
（3）java.util 包含实用工具类，如 Date，Scanner 类。
（4）java.net 包含与网络通信相关的类。
……

4.3　Create packages（创建包）

As a rule, putting each Java source file into a package, you need to create packages first to store your Java source files and bytecode files. If you never create a package in a Java project, a default package (without name) is provided for you, all the source and bytecode files are stored in it. In Eclipse, select a Java project, then New → Package, you'll enter into the "New Java Package" dialogue, at the Name textbox, enter a package name. Dots are used to separate package names. For instance p1.p2, the dot separates the super package p1 from its subpackage p2. The New Java Package dialogue is shown in Figure 4-3. A dot is equivalent of "/", "\", to separate a folder from its subfolders.

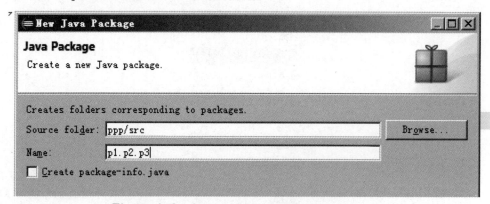

Figure 4-3　Java package creating dialogue

遵循每个 Java 源文件都在某个包里的规则，在创建类之前要先创建包，用以存放 Java 源文件和字节码文件。如果没在 Java 项目里创建包，系统会提供一个默认包（无名包），将所有源文件和字节码文件都放入其中。在 Eclipse 环境下，选中一个 Java 项目，通过 New → Package 进入 New Java Package 对话框，在 Name 文本框里输入包名。用 "." 分隔子包名。例如 p1.p2，"." 将包 p1 与其子包 p2 隔开。New Java Package 对话框如图 4-3 所示。"." 与文件夹名里的 "/" 或 "\" 将文件夹与子文件夹分隔的作用一样。

In fact, each package produces two subfolders which take the same package name as their folder names, one is under "src" folder and another is under "bin" or subfolders of "src" and "bin" respectively. The folders under "src" store .java files and "bin" for .class files. Figure 4-4 shows the package structure of a Java project and figure 4-5 shows the corresponding folder structure. Under the Eclipse environment, creating a Java class in the Create New Class dialog, the package statement will be putted into the source file as the first statement automatically. In example 4-1, the first statement "package org;" is inserted automatically while class Calculator is created in the Eclipse environment.

事实上，创建一个 Java 包会分别在 src 和 bin 文件夹下或者它们的子文件夹下生成与包名相同的两个子文件夹，src 之下的文件夹用于存放 .java 文件，bin 用于存放 .class 文件。图 4-4 是一个 Java 项目的包结构，图 4-5 是与之对应的文件目录结构。在 Eclipse 环境下，在创建新类对话框里创建一个 Java 类时，包语句会被自动地加入这个类所在的源文件，成为第一条语句。例 4-1 的第一条语句 package org; 是被自动加入的。

Figure 4-4　Package structure in Eclipse　　Figure 4-5　File directory structure

【Example 4-1】 **Define a class in package org**

```
package org;
public class Calculator {
  public int add(int x, int y) {
    return(x+y) ;
  }
}
```

【Example 4-2】 **Define another class in package mypack.sub1.sub2**

```
//only comments can be here
package mypack.sub1.sub2; //first line
public class HelloWorld {
  public static void main(String[] args) {
    System.out.println("Hello World");
  }
}
```

4.4 Import packages(导入包)

The purpose to organize classes and interfaces with packages is for using them conveniently. There are two ways to use the public classes stored in a package.

用包组织类和接口的目的是为了方便地使用类和接口。有两种方法使用包里的 public 类。

1. Use the full class name (include package name)

All built-in classes are grouped into several sub packages under the "java" package. Package names can be used to identify the class you want to use.

In example 4-3, class Date is used with its full name "java.util.Date()".

系统提供的类都在 java 包的各个子包内,例 4-3 用类的全称使用 Date 类。

【Example 4-3】**Class under "java" package**

```java
public class HelloDate {
  public static void main(String[] args) {
    System.out.println("Hello, it's: ");
    System.out.println(new java.util.Date());
  }
}
```

2. Use import keyword

If you need to use other classes not included in package java.lang, you must tell Java compiler exactly what classes and which packages the classes are by the import keyword. An import statement tells the compiler to bring in a class or all classes in a specific package. However, a default package can't be imported to any java source files for lack of package name, so we need to create packages for classes storing and using (importing). The import statements must next to the package statement.

如果需要用 java.lang 包以外的类,必须利用 import 关键字准确地告诉编译器所需的类,以及包含该类的包。import 语句通知编译器将指定包里的一个或所有类导入当前文件。然而,默认包由于没有名字而无法导入任何 Java 源文件,所以要创建包,用于类的存储与使用(导入)。import 语句必须在 package 语句之后,其他语句之前。

Import a specific package member: for instance import mypackage.Calculator;

Or import an entire package: for instance import mypackage.*;

Using "*", the file will be compiled more slowly. But "*" does not affect the performance and the size of the class.

Example 4-4 uses Date class through importing the class "java.util.Date" in the source file. Example 4-5 and example 4-6 create classes under package world.moon, and in example 4-7, the classes created in package world.moon are imported into the file and are used directly by the class names.

【Example 4-4】 **Use Date class through importing the class "java.util.Date" in the source file**

```
import java.util.Date; //import java.util.*;
public class HelloDate {
  public static void main(String[] args) {
    System.out.println("Hello, it's: ");
    System.out.println(new Date());
  }
}
```

【Example 4-5】 **Create a class HelloMoon under package world.moon**

```
package world.moon;
public class HelloMoon {
  private String holeName;
  public String getHoleName() {return holeName;}
  public HelloMoon(String hName) {
    holeName=hName;
  }
}
```

【Example 4-6】 **Create another class My under package world.moon**

```
package world.moon;
public class My {
  public My() {
     System.out.println("MyPackage!");
  }
}
```

【Example 4-7】 **In package org.p1.p2, create objects of classes m in the package world.moon**

```
package org.p1.p2;
import world.moon.*; //next line behind the package statement
public class For {
  public static void main(String[] args) {
    HelloMoon hm=new HelloMoon("rabbit hole!");
    My m=new My();
       System.out.println(hm.getHoleName());
  }
}
```

Note

We can use classes in a package by importing the package in the present source file, but can't use any classes that belong to its sub packages. For example, classes Scanner and Date are in package java.util, util is the sub package of java, you can only import java.util.*, instead of import java.*.

注意

通过import语句导入包名时,不能使用该包的子包里的类。例如Scanner和Date类在java.util里,util是java包的子包,你只能import java.util.*而不能import java.*。

If different packages contain the same class name, you need to identify them by package names.

For instance:

```
import tom.jiafei.*; //contains class AA
import sun.com.*; //contains class AA
tom.jiafei.AA a=new tom.jiafei.AA();
sun.com.AA b=new sun.com.AA();
```

4.5 Package java.lang(java.lang包)

Package Java.lang contains all language related classes. Figure 4-6 shows several classes in java.lang.

Java.lang is imported implicitly automatically in every Java source file, so never use the import keyword to import java.lang in a java source file.

java.lang包含了与Java语言相关的类,包里部分类如图4-6所示。java.lang包被自动地加入每个Java源文件,因此,不需要导入java.lang包。

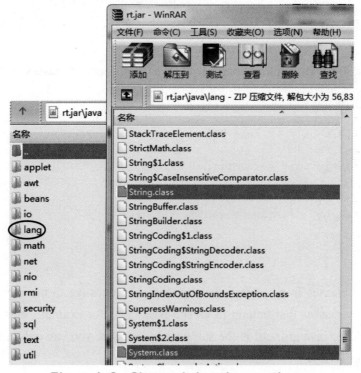

Figure 4-6 Classes in java.lang package

4.6 Useful classes in package java.lang（java.lang 包中常用的类）

Package java.lang contains a collection of language related classes.

4.6.1 Object class and its toString method（Object 类和它的 toString 方法）

Object is the root class of all classes. It has several frequently used methods, some of them we've already used in our java source files, especially the toString(). Method toString() is automatically called in special situations when the compiler needs a String but an object there. The prototype of to toString() is as bellow:

public String toString()

Its return type is String. The toString() returns: class name+@+Hash address of an object.

Object 类是所有类的根类。它有很多常用的方法，有些已经在 Java 源文件里使用了，特别是 toString() 方法。当需要 String 类型，而实际类型是对象时，toString() 会被自动地调用。

toString() 的原型是：public String toString()

它的返回类型是 String，toString() 返回：类名 +@+ 对象的 Hash 地址。

【Example 4-8】Use method toString()

```
public class Point {
  private int x,y;
  public Point(int a,int b) {
    x=a;
    y=b;
  }
  public static void main (String[] args) {
    Point x=new Point(4,3);
    System.out.println(x); //or System.out.println(x.toString());
  }
}
```

Output:

Point@182da3d

Where, method println needs a String argument, but the actual parameter x is an object, so the method toString() is invoked automatically and outputs class name Point and the Hash address of x. At times, we don't care about an object's address, but special things, like coordinates of a point, in that case, just override (different from overload) the toString().

其中，println 需要 String 参数，而实际参数是对象 x，因此 toString() 被自动地调用，输出类名 Point 和对象的哈希地址。有时希望输出特定的内容，例如点的坐标而不是对象的地址，用户可以重写（与重载不同）toString()。

Insert codes into example 4-8 as bellow:

```
public String toString() {
  return "point:"+x+","+y;
}
```

You'll see the output turns out:

point:4,3

4.6.2 System class（系统类）

Class System in java.lang package has several object fields, the most of famous two of them are in and out. The out is a PrintStream type object. Class PrintStream has a list of methods you can call, such as methods print and println.

For instance:

```
System.out.println("Hello World!");
```

System 类里有几个对象 fields，其中最有名的两个是 in 和 out。Out 是 PrintStream 类型的对象。PrintStream 类提供了很多方法，例如方法 print 和 println。

Exercises

1. Create packages.

(1) Create two packages: pkg1 and pkg2 in a Java project.

(2) Under pkg1, create two sub-packages: bag1 and bag2.

创建包

(1) 在一个 Java 项目下创建 2 个包：pkg1 和 pkg2。

(2) 在 pkg1 下再创建 2 个包：bag1 和 bag2。

2. Override toString() method.

(1) Read and run the program below.

```
public class Student {
  private int sId;
  private String sName;
  private static int sTotalNo;
  public Student(int id,String name) {
    sTotalNo++;
    sId=id;
    sName=name;
  }
  public static void main(String[] args) {
    Student s1=new Student(101,"Jeff Bissell");
    Student s2=new Student(102,"David Trump");
    System.out.println(s1);
    System.out.println(s2);
    System.out.println("Total number "+sTotalNo);
```

    }
  }

(2) Override toString() to change the output to:

StudentID: 101 Name: Jeff Bissell

StudentID: 102 Name: David Trump

Total number 2

3. Complete the program below.

(1) Override toString() method, let it output: "坐标：x=12, y=36".

(2) Define method main. In main, create an object, call method move, and print the result by using println(object).

```
public class Point {
  private int x, y;
  public void move(int a, int b) {
    x=a;
    y=b;
  }
  //override toString here
  //main method here
}
```

完成下面的程序：

(1) 重写 toString() 方法，使输出结果类似"坐标：x=12, y=36"。

(2) 定义 main 方法，在其中创建一个对象，调用 move 方法，再用 println(对象) 输出结果。

Chapter 5
Inheritance（继承）

In prior chapters, we have learnt two Java OOP properties: Encapsulation (or Abstraction) and Polymorphism. Inheritance is another important property of OOP.

5.1 What is inheritance?（什么是继承？）

Some classes have common features. The best way to design these classes to avoid redundancy is to reuse the existing classes, known as inheritance. The class that inherits an existing class is called derived or inherited class and the original calss is called base class.

A derived class automatically has all the instance fields and all the methods of its base class, and can have additional methods and additional instance fields. When you define a derived class, you add new instance fields and new methods.

Java does not support multi inheritance. A class can only extend one base class. In C++, a derived class can derive from more than one base class. Figure 5-1 shows the two different inheritance mechanisms. However Java supports implementing multiple interfaces, which is similar to the multiple inheritance of C++.

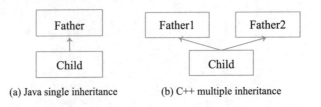

Figure 5-1 Inheritance comparison of C++ with Java

When relationship exists between two classes, we use inheritance. The parent class is termed super or base class and the inherited class is the subclass. The extends keyword is used by the subclass to inherit the super class.

Chapter 5　Inheritance（继承）

有些类具有相同的特性，为了避免重复设计，最好的办法是重用已经存在的类，这种做法称为继承。继承下来的类称为派生类或继承类。

派生类自动继承基类的成员变量和方法。定义派生类时，只需加入新的成员变量和方法。Java 不支持多重继承，一个派生类只能有一个基类。而 C++，一个派生类可以有多个基类。然而 Java 支持接口的多重实现，类似于 C++ 的多重继承。

当类之间有关系时，就使用继承。被继承的类叫作 super 类或者 base 类，继承类称为子类。关键字 extends 用于子类继承 super 类。

For example:

```
class Person {
   //fields and methods
}
class Student extends Person {
   /*inherit the fields and methods of Person and in addition add its own specialties*/
}
```

5.1.1　Root class Object（根类对象）

You are always doing inheritance when you define a class, unless you explicitly inherit from some other classes, you implicitly inherit from Java's standard root class Object. Every class in Java is descended from class Object. If no inheritance is specified when a class is defined, its super class is Object. Object class exists in Java JRE's class library within package java.lang. Package java.lang is inserted into every Jave source file automatically. Figure 5-2 shows class Account inherits all methods from Object. Figure 5-3 shows usable methods of Object. Actually, a class inherits all fields and methods of Object, includes unseen members.

如果定义类时未显式地继承其他类，其实已经在继承类了，隐式地继承了 Java 的根类对象。Java 的所有类都派生于 java.lang 包里的 Object 类。定义类时，如果没有继承其他类，都以 Object 为 super 类。Object 类在 JRE 类库的 java.lang 包里，java.lang 包被自动地插入到每个 Java 源程序。图 5-2 显示所有从 Object 类继承的方法。图 5-3 显示 Object 类的可用方法。事实上，一个类会继承 Object 的全部成员变量和方法，包括不可见成员。

```
public class Account {
   public static void main(String[] args) {
      Account person=new Account();
```

Figure 5-2　Callable methods of Account　　Figure 5-3　Callable methods of Object

Note that Account class has all nine methods defined in Object. That is: class Account {…} equivalent to class Account extends Object {…}.

5.1.2 Defining subclasses（定义子类）

Defining a subclass, use the keyword followed by the name of the base class. When you do this, you automatically get all the fields and methods in the base class. You can also add new fields, add new methods, override methods of the super class.

Overriding a method is different from overloading, the method is overridden stays at its base class (direct or indirect), but not in the current class.

定义子类时，使用关键字 extends，其后跟基类的类名。子类继承基类的全部成员变量和方法，你还可以加入新的成员变量，加入新的方法，重写（覆盖）基类（直接或间接基类）的方法。

方法重写不同于方法重载，被覆盖（重写）的方法在基类，而不在当前类。

【Example 5-1】**Super and subclass definitions**

```java
class Student {                              //super class
  public int stu_no;
  public String stu_name;
  void set_no(int no) {
    stu_no=no;
  }
  void set_name(String name) {
    stu_name=name;
  }
  public String toString() {
    return "学号:"+stu_no+" 姓名:"+stu_name;
  }
}
public class Graduate extends Student{   //subclass
  public String supervisor;
  void set_super(String name) {
    supervisor=name;
  }
  public static void main(String[] args) {
    Graduate stu=new Graduate();
    stu.set_no(100);
    stu.set_name("刘美丽");
    stu.set_super("王爱国");
    System.out.print(stu);
    System.out.print(" 导师:"+stu.supervisor);
  }
}
```

Output:

学号:100 姓名:刘美丽 导师:王爱国

The subclass adds only additional fields and methods.

【Example 5-2】 **Add addtional fields and methods in the subclasses**

```
class ASuper {
  private int x=1, y=2;
  public void outxy() {
    System.out.print("x="+x+" y="+y+" ");
  }
}
class BSuper extends ASuper {
  private int z=3;
  public void outz() {
    System.out.print("z="+z+" ");
  }
}
class C extends BSuper {
  private int u=4, v=5;
  public void outuv() {
    System.out.print("u="+u+" v="+v);
  }
}
public class Test {
  public static void main(String[] args ) {
    ASuper a=new ASuper();
    BSuper b=new BSuper();
    C c=new C();
    c.outxy();
    c.outz();
    c.outuv();
  }
}
```

Output:

x=1 y=2

z=3

u=4 v=5

Figure 5-4 shows all the visible (callable) members of object "c" of class C. The five fields x, y, z, u and v are private members of object "c" and is unseen outside the classes where they are defined. The private members of other classes can't be referenced by object "c" defined in class Test, so they are invisible in the member list of object "c". Change all five fields from private to default or to public, some of the visible members of object "c" are as shown in Figure 5-5. All the members in the lists of Figure 5-4 and 5-5 can be referenced by "c", through the class member select operator ".".

图 5-4 列出了对象 c 的所有可见成员。5 个 fields（数据成员）x, y, z, u 和 v 是对象 c 的私有成员，在定义它们的类外是不可见的。在 Test 类里定义的对象 c 不能访问在其他类里定义的私有成员，因此它们没有出现在对象 c 的可访问成员列表。图 5-4 中，将 5 个数据成员的访问

属性都改成默认的或者 public 的之后，对象 c 的部分可见成员列表如图 5-5 所示。图 5-4 和 5-5 中列出的所有成员都可以通过对象 c 和类的成员选择运算符 "." 访问。

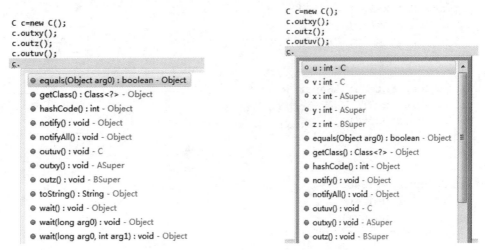

Figure 5-4　All visible members of object c　　Figure 5-5　Some visible members of object c

5.2　Super keyword（super 关键字）

Usually, an object of a subclass owns much memory than its super class's object. The fields of a subclass are equal to fields defined in the super class plus fields defined in the subclass, so the subclass needs more memory. When an object of the subclass is created, the constructors, both of the super and the subclasses are invoked to initialize the part of the fields defined in their own class respectively. Figure 5-6 shows the initialization rule.

通常一个子类对象拥有的存储空间比它的基类对象的存储空间大。子类对象所拥有的存储空间 = 基类的 fields+ 子类的 fields 所需的存储空间。创建子类对象时，基类和子类的构造方法都被调用，分别负责在自己类里定义的那部分 fields 的初始化，如图 5-6 所示。

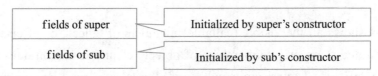

Figure 5-6　Constructors initialize their own fields

5.2.1　Super corresponding to default constructor（默认构造方法的 super）

Super classes' constructors (direct or indirect super classes) are not inherited. They are invoked explicitly or implicitly. Using the super keyword, they will be invoked explicitly. A constructor is used to construct an object of a class and initializes the object. A super class' constructor is invoked in the subclasses' constructors. The super() corresponds a super class' default constructor and must be the first statement in a subclass' constructor. A super class' constructor is always invoked. Even if super() is not involved, the compiler puts super() to the subclass' constructor implicitly. When creating an object (instance) of a class, all its super classes' constructors along the inheritance chain will be invoked. In example 5-3, the super(); are inserted into the subclasses' constructors, Drawing and Cartoon, in the hidden way and represent constructors of Art and Drawing respectively.

子类不继承基类（直接或间接基类）的构造方法。基类的构造方法或显式或隐式地被调用。使用 super 关键字可以显式地调用基类的构造方法。构造方法用于创建类的对象与初始化对象。基类的构造方法在子类的构造方法里调用。super() 代表基类的缺省（无参）构造方法，它必须是子类构造方法的第一条语句。基类的构造方法在任何情况下都会被调用。如果子类的构造方法不含 super()，super() 会被自动地加入。创建类的对象（实例）时，所有基类的构造方法都会按继承顺序被一个个地调用。在例 5-3 中，super(); 被隐式地插入到 Drawing 和 Cartoon 的构造方法中，分别对应 Art 和 Drawing 的构造方法。

Suppose Parent is a class.

```
class SuperTest extends Parent {      equivalent to    class SuperTest extends Parent {
  public SuperTest() {                                   public SuperTest() {
    …                                                      super();
  }                                                        …
  …                                                      }
}                                                        …
                                                       }
```

【**Example 5-3**】**The super(); are inserted into the subclasses' constructors in the hidden way**

```
class Art {
  Art() {
    System.out.println("Art constructor");
  }
}
class Drawing extends Art {         super(); //represents Art's constructor
  Drawing() {
    System.out.println("Drawing constructor");
  }
}
public class Cartoon extends Drawing {   super(); //represents Drawing's constructor
  public Cartoon() {
    System.out.println("Cartoon constructor");
  }
  public static void main(String[] args) {
    Cartoon x=new Cartoon();
  }
}
```

Art
↑
Drawing
↑
Carton

Output:

Art constructor
Drawing constructor
Cartoon constructor

Example 5-1, here, example 5-4, both of the super class and the subclass have no constructors, so their default constructors are provided and invoked.

【Example 5-4】 **Call the default constructor in super and subclass**

```
class Student { //super class
  private int stu_no;
  private String stu_name;
  void set_no(int no) {
    stu_no=no;
  }
  void set_name(String name) {
    stu_name=name;
  }
  public String toString() {
    return "学号:"+stu_no+" 姓名:"+stu_name;
  }
}
public class Graduate extends Student { //subclass
  private String supervisor;
  void set_super(String name) {
    supervisor=name;
  }
  public static void main(String[] args) {
    Graduate stu=new Graduate();
    stu.set_no(100);
    stu.set_name("刘美丽");
    stu.set_super("王爱国");
    System.out.print(stu);              //System.out.print(stu.toString);
    System.out.print(" 导师:"+stu.supervisor);
  }
}
```

 Output:

学号:100 姓名:刘美丽 导师:王爱国

5.2.2 Super corresponding to constructors with arguments
（有参构造方法的 super）

If you want to call a super class' constructor that has arguments, you must explicitly call the base class' constructor using the super with an appropriate argument list. Now we come to the conclusion: each constructor of a subclass has a super as its first statement.

如果基类的构造方法含参数，必须在子类的构造方法里使用带参数的 super。因此，所有子类的构造方法的第一条语句都是 super。

【Example 5-5】**Call the super class'constructor with arguments explicitly in the subclass**

```
class Game {
  Game(int i) {System.out.println("Game constructor "+i);}
}
class BGame extends Game {
  BGame(int i) {
    super(i);
    System.out.println("BGame constructor");
  }
}
public class Chess extends BGame {
  Chess(int k) {
    super(k);
    System.out.println("Chess constructor");
  }
  public static void main(String[] args) {
    Chess x=new Chess(11);
  }
}
```

Output:

Game constructor 11

BGame constructor

Chess constructor

【Example 5-6】**Mixed use of base class constructors in the subclasses**

```
public class Faculty extends Employee {
  public static void main(String[] args) {new Faculty();}
  public Faculty() {
    super("Employee");
    System.out.println("Faculty");
  }
}
class Employee extends Person {
  public Employee(String s) {
    System.out.println(s);
  }
}
class Person {
  public Person() {
    System.out.println("Person");
  }
}
```

Output:

Person
Employee
Faculty

5.3 Order of constructor calls（构造方法的调用次序）

The order of constructor calls is as such:
(1) The base classes' constructors in order of inheritance chain.
(2) The object fields' constructors in order of declaration.
(3) The object's own constructor at last.
构造方法的调用次序是：
（1）按继承次序，调用 base 类的构造方法；
（2）按类的对象成员的声明次序调用对象成员的构造方法；
（3）最后调用对象自己的构造方法。

【Example 5-7】Example of constructor call

```java
class Meal {
  Meal() {
    System.out.println("Meal");
  }
}
class Bread {
  Bread() {
    System.out.println("Bread");
  }
}
class Cheese {
  Cheese() {
    System.out.println("Cheese");
  }
}
class Lunch extends Meal {
  Lunch() {
    System.out.println("Lunch");
  }
}
class PtLunch extends Lunch {
  PtLunch() {
    System.out.println("PortableLunch");
  }
}
```

```
public class SandWich extends PtLunch {
  private Bread b=new Bread();
  private Cheese c=new Cheese();
  public SandWich() {
    System.out.println("SandWich");
  }
  public static void main(String[] args) {
    new SandWich();
  }
}
```

 Output:

Meal

Lunch

PortableLunch

Bread

Cheese

SandWich

【Example 5-8】 **Object fields**

```
public class Test {
  int x;
  QQ s=new QQ(120);
  public Test(int a,int b) {
    x=a;
    s.u=b;
    System.out.println("I'm Test.x="+x+",u="+b);
  }
  public static void main(String args[]) {
    SubTest ex=new SubTest(12,15,33);
    ex.output();
  }
}
class SubTest extends Test {
  private QQ pp=new QQ(12);
  public SubTest(int p,int q,int r) {
    super(p,q);
    pp.u=r;
    System.out.println("I'm SubTest.u="+pp.u);
  }
  public void output() {
    System.out.println("x="+x+",s.u="+s.u+",pp.u="+pp.u);
  }
}
class QQ {
  int u;
  public QQ(int a) {
```

```
        u=a;
        System.out.println("I'm QQ.u="+u);
    }
}
```

Output:

I'm QQ. u=120
I'm Test. x=12,u=15
I'm QQ. u=12
I'm SubTest. u=33
x=12,s.u=15,pp.u=33

5.4 Final keyword（final 关键字）

Final means never change, is used to define constants. The final keyword can be used at fields, methods and classes. A value should be given at the time of defining a constant.

```
final int MAX=2000;
```

5.4.1 Final fields（final 成员）

Fields that are both static and final (compile-time constants) are static constants, by convention, they are capitalized and use underscores to separate words.

A value must be assigned to at the time of defining a static constant. However for a non-static constant, you can perform the assignment to the final field either at the point of the field definition or in the constructor. In example 5-9, all the non-static final fields, ValueOne, v2 and ak, are all assigned at the defining time. A non-static final field is a constant within an object and each object can have its own constant value. We call it the blank final while define a non-static final field without assigning it a value. In example 5-10, jk and ak are the blank final, and for object bf and cf, final jk has different values. If a non-static final field is assigned a value at defining time, you can't assign it in a constructor again, only one apporturnity to assign a non-static final field, at defining time or in a constructor.

Final references rather than primitive fields, easily cause confusing. With a primitive field, final makes the value a constant, but with a reference, final makes the reference constant. Once an object is assigned to a reference, the reference can never be changed to point to another object. However, the object itself can be modified. This restriction includes arrays which are also objects.

同时用 static 和 final 修饰的 fields 是静态常量，按照常规要大写，单词用下画线分开。

定义静态常量时必须赋值，而对于非静态常量，既可以在定义时赋值，也可以在构造方法里赋值。例 5-9 的 3 个非静态常量 ValueOne，v2 和 ak 都在定义时被赋了值。非静态常量是每个对象里的常量，不同对象可以有不同的常量值。把在定义时未赋值的非静态常量称为空常量。例 5-10 里，jk 和 ak 都是空常量，jk 在对象 bf 和 cf 里取不同的常量值。如果在定义时为非静态常量赋了值，就不允许在构造方法里再次为其赋值。二者只能取其一。

当 final 修饰的不是基本数据类型，而是引用类型时，容易引起混淆。用 final 修饰的基本数据类型是常数，但修饰引用时，引用是常数，它只能指向初始化时指向的对象，不能指向其他对象，但对象本身是可以改变的。这种限制也适用于数组，因为数组也是对象。

【Example 5-9】**Difference of final and non-final fields**

```
class Value {
  int i;
  public Value(int is) {i=is;}
}
public class FinalData {
  private String id;
  public FinalData(String sid) {id=sid;}
  private final int ValueOne=9;              //non-static final
  private static final int V_TWO=99;
  private Value v1=new Value(11);
  private final Value v2=new Value(22);      //non static final
  private static final Value V3=new Value(33);
  private final int[] ak={1,2,3,4,5,6};      //array, non-static final
  public static void main(String[] args) {
    FinalData fd1=new FinalData("fd1");
    fd1.v2.i++;
    fd1.v1=new Value(12);                    //okay
    System.out.print("non-static final array: ");
    for(int i=0; i<fd1.ak.length; i++) {
      fd1.ak[i]++;                           //can't fd1.ak=new int[3];
      System.out.print(fd1.ak[i]+",");
    }
    System.out.println("\nstatic final V_TWO: "+V_TWO);
    FinalData fd2=new FinalData("fd2");
    System.out.println("non-static final ValueOne of fd2: "+fd2.ValueOne);
  }
}
```

Output:

non-static final array: 2,3,4,5,6,7,

static final V_TWO: 99

non-static final ValueOne of fd2: 9

【Example 5-10】**Assignment of final and non-final fields**

```
class A { int a;}
class BlankFinal {
  final int i=0;
  final int jk; //blank final field
```

```
    final A ak;    //blank reference final
    BlankFinal() {
      jk=1;    //must assign a value in the constructor
      ak=new A();
    }
    BlankFinal(int x) {
      jk=x;    //must assign a value in the constructor
      ak=new A();
    }
    public static void main(String args[]) {
      BlankFinal bf=new BlankFinal(12);
      BlankFinal cf=new BlankFinal(76);
      System.out.println("non-static final jk of bf: "+bf.jk);
      System.out.println("non-static final jk of cf: "+cf.jk);
    }
  }
```

Output:

non-static final jk of bf: 12
non-static final jk of cf: 76

5.4.2　Final arguments（常参数）

Java allows you to make arguments final by declaring them as such in the argument list. This means that inside the method you cannot change the final arguments.

```
void test(final int i) {
  i++; //error, can't change
}
```

5.4.3　Final methods（常方法）

If a method is set to final, means you put a "lock" on the method to prevent any inheriting class from changing, forbid overriding.

If a class is final means you can't inherit from the class or can't allow anyone else to do so.

It prevents inheritance, so methods in a final class are implicitly final, since there's no way to override them.

如果一个方法是 final 的，相当于给这个方法加锁，防止继承类去重写该方法，确保该方法在继承的过程中保持不变且不被覆盖。

一个 final 类不能被继承，它所有的方法也隐含地被声明为 final，既然不能继承，当然无法重写。

```
public final class AFinal {
  …
}
```

5.5　Access specifiers（访问说明符）

There are four types of access specifiers for protecting fields and methods:

(1) private: Accessible only within a class.
(2) default (no specifier): Accessible only within a package.
(3) protected: Accessible within a package and all its child classes in different packages.
(4) public: Accessible everywhere.

A class can only be public or default (no modifier).

有 4 种用于保护类的成员变量和方法的修饰符：

（1） private: 类内可以访问。

（2） default (no modifier): 缺省（无访问说明符），包内可以访问。

（3） protected: 包内可以访问，其在不同包里的子类可以访问。

（4） public: 处处可以访问。

类只能用 public 说明，或者不被任何说明符说明。

A member without any specifier (sometimes call it "friendly" member) is referred to as package access. It means that all the other classes in the current package can access to that member, but to all the classes outside of this package, the member appears to be private. Since a compilation unit, a file, belongs only to a single package, all the classes within a single compilation unit are automatically available to each other via package access. Package access allows you to group related classes together in a package so that they can easily interact with each other via package access.

The protected specifier is similar to the default, also available to all child classes. The inherited classes in another packages can access to protected members of their father.

The order of increasing access level: private → default → protected → public.

Table 5-1 lists the access restrictions.

没有修饰符的成员在包内可以访问（有时称其为友好成员）。当前包的其他类都可以访问它，而对于当前包之外的类，该成员是包内私有，不能被访问。一个文件，即一个编译单元属于一个包。通过包访问限制，包内成员可以互相访问，有利于将有关系的类组织到一个包里，而对外禁止访问。

用 protected 修饰的成员在包内可以访问，包外的子类也可以访问。包外的子类可以访问其基类的保护的成员，即使子类与其基类不在同一个包里。

Table 5-1　Access specifiers

Modifier	Applicable to	Who can access
private	fields and methods	all members within the same class
no modifier	fields, methods, classes and interfaces	all classes in the same package
protected	fields and methods	all classes in the same package and all subclasses, even they residing in a different package
public	fields, methods, classes and interfaces	any class

Example 5-11 is an example of the private modifier. Remember: never touch private members outside that class!

【Example 5-11】 Use the "private" modifier

```
class Sun {
  private Sun() {}
  static Sun makeASun() {
    return new Sun();     //create an object
  }
}
public class IceCream {
  public static void main(String[] args) {
    Sun x=new Sun();      //error
    Sun x=Sun.makeASun ();
  }
}
```

A constructor can be specified by any one of the four specifiers, which is different from C++.

Make a member public, and then every member, everywhere they can be accessed to. Example 5-12 shows public classes, HelloMoon and My, defined in package world.moon, can be used in package diff. The following three classes are defined in three different Java files.

【Example 5-12】 Public access in different packages

```
package world.moon;
public class HelloMoon {
  private String holeName="rabbit hole";
  public String getHoleName() {return holeName;}
  public HelloMoon(String hName) {
    holeName=hName;
  }
}

package world.moon;
public class My {
  public My() {
    System.out.println("MyPackage!");
  }
}

package diff;
import world.moon.*;
public class For {
  public static void main(String[] args) {
    HelloMoon hm=new HelloMoon("rabbit");
    System.out.println(hm.getHoleName());
    My m=new My();
  }
}
```

【Example 5-13】Package access

```
package world.moon;
class HelloMoon {
  private String holeName="rabbit hole";
  public String getHoleName() {
    return holeName;
  }
  public HelloMoon(String hName) {
    holeName=hName;
  }
}
package world;
class My {
  public My() {
    System.out.println("MyPackage!");
  }
}
package theother;                          //in default package
import world.*;
import world.moon.*;
public class For {
  public static void main(String[] args) {
    HelloMoon hm=new HelloMoon("A hole");  //error
    My m=new My();                         //error
    System.out.println(hm.getHoleName());
  }
}
```

Example 5-13, trying to access to the non-public classes HelloMoon and My in package theother causes errors. Change classes HelloMoon and My to public, the above two statements marked with "error" will be right.

【Example 5-14】Protected access

```
package package1;
public class Father {
  private int p1=1;
  int p2=2;              //default
  protected int p3=3;
  public int p4=4;
  public int retp1() {
    return p1;      //p1 private
  }
  public int retp2() {
    return p2;      //p2 default
  }
}
package package1;
public class A {
  void usep3() {
```

```
      Father a=new Father();
      System.out.println(a.p2);
      System.out.println(a.p3); //p3 protected
    }
}
package package2;
import package1.*;
class Test {
  public static void main(String[] args) {
    Father f=new Father();
    Son s=new Son();
    System.out.print("pri="+f.retp1()+",def="+f.retp2()+",prot="+
    s.retp3()+",pub="+f.p4);
  }
}
class Son extends Father {
  int retp3() {
    Son f=new Son();
    f.p3=3;
    Father fat=new Father();
    fat.p3=2;          //error
    return p3;
  }
}
```

Output:

pri=1,def=2,prot=3,pub=4

【Example 5-15】**Access application**

```
public class Protection {
  private int p1=1;
  int p2=2;                              //default
  protected int p3=3;
  public int p4=4;
  public void fun() {
    System.out.println(p1);
    System.out.println(p2);
    System.out.println(p3);
    System.out.println(p4);
  }
  int retp1() {return p1;}
}
class Derived extends Protection {
  public void function() {
    System.out.println("private="+p1);   //error
    System.out.println("default="+p2);
    System.out.println("protected="+p3);
    System.out.println("public="+p4);
    System.out.println("public="+retp1());
```

```
    }
  }
  class A {
    public void method() {
      Derived t=new Derived();
      System.out.println("default="+p2);         //error
      System.out.println("protected="+p3);       //error
      System.out.println("public="+p4);          //error
      System.out.println("default="+t.p2);
      System.out.println("protected="+t.p3);
      System.out.println("public="+t.p4);
    }
  }
```

5.6 Polymorphism(多态)

Polymorphism is the third essential feature of the object oriented programming (encapsulation, inheritance). Polymorphism refers to "Many forms". Polymorphism accords with the thinking way of people, that is for a given command (a given method call), but objects respond differently although they are closely related. Behaves or actions of objects are realized through methods, so calling a method, equivalent to giving a command to start a task. A given command triggers different actions. Polymorphism supports building extensible systems, which is essential for software developing. There two types of polymorphisms, static and dynamic. We've already learnt the static polymorphism. Example 5-16 realizes the static polymorphism.

多态是面向对象程序设计的第三个重要特性(封装、继承),多态可以看成是多种表现形式。对于同一命令(调用同一个方法),多态符合人们的思维方式,不同的相关的对象产生不同的行为。对象的行为或反应是通过方法实现的,因此调用方法,与发出一个开始行动的命令效果相同。多态是指不同对象,对于同一命令,做出不同的响应的能力。多态支持可扩展系统的开发,这是软件开发的基本要求。有两种形式的多态,静态与动态。我们已经学习了静态多态,例5-16实现了静态多态。

【Example 5-16】Example of the static polymorphism

```
class OverLoading {//static polymorphism
  void plan() {
    System.out.println("no arguments");
  }
  void plan(int x) {
    System.out.println("The value of x is "+x);
  }
}
public class OverLoad {
  public static void main(String args[]) {
    OverLoading ov=new OverLoading();
```

```
        ov.plan();
        ov.plan(5);
    }
}
```

Where, which plan is called is resolved at the compile time.

How to realize the dynamic polymorphism?

Answer: For classes related by inheritance, design different implementations for a given method in its subclasses.

In order to realize the dynamic polymorphism, the methods should possess the following two basic properties:

(1) Same: A given method (same signature, such as name, arguments, return type)

(2) Different: Overriding the given method (different behaves) in derived subclasses.

例中哪个 plan 被调用是在代码编译时确定的。

如何实现动态多态呢？对于具有继承关系的类，对于指定的同一个方法，在其子类给出不同的实现。实现动态多态，特定的方法需具备以下两个特点：

（1）相同点：同一个方法（名字、形参和返回类型相同）。

（2）不同点：在子类里重写其父类里的该方法。

5.6.1　Method overriding（方法重写）

Sometimes it is necessary for the subclass to modify the implementation of a method defined in the super class. This is referred to as method overriding.

We've overridden the toString() several times, for instance:

```
public String toString() {
    return super.toString()+" radius is "+radius;
}
```

Figure 5-7 shows a method overriding example. Method treatPatient() is overridden in subclasses Surgeon and ChineseDoc, and keeps unchanged in subclass FamilyDoc.

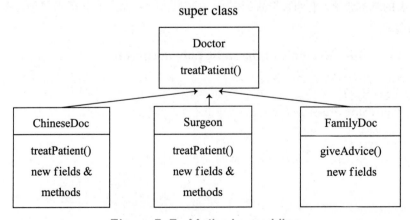

Figure 5-7　Method overriding

5.6.2 Upcasting and dynamic polymorphism（升级转换与动态多态）

We can say a subclass is a new type of its super class. Consider a super class called Fruit, Apple and Orange are its subclasses. An apple is a fruit, so you can always safely assign an Apple object to a Fruit reference. However, a fruit is not necessarily an apple. A super class can reference to subclass' objects. A Surgeon's object is always an object of Doctor, but a Doctor object is not necessarily an object of Surgeon, for instance:

```
Doctor surgdoc=new Surgeon(); //surgeon is a doctor
Doctor doctor=new Doctor();   //doctor not necessarily a surgeon
```

可以把子类看成是由基类派生出的一个新类型。以基类水果为例，苹果和橘子是其子类。苹果当然是水果，因此可以把一个苹果对象赋值给水果引用。当然，水果不必是苹果。基类（的引用）可以指向子类的对象。外科医生（Surgeon）对象可以作为医生（Doctor）的对象，即可以将外科医生对象赋值给医生的引用，反之医生对象不能赋值给外科医生的引用。

```
Doctor surgdoc=new Surgeon(); //surgeon is a doctor
surgdoc.treatPatient();
```

Which version of the methods is invoked? It invokes Surgeon's treatPatient() method, because the reference is pointing to a Surgeon's object. If a base class' reference is used to invoke a method, the method to be invoked is decided at run time, depending on the object (instance) that the reference is referencing to, not the reference itself.

Although surgdoc is a reference of Doctor, it invokes the method of Surgeon, as it points to a Surgeon's object, which is decided during the run time and termed dynamic or run-time binding, that is dynamic polymorphism. The dynamic polymorphism is achieved by a generalized reference and a specific object (instance). In chapter three, we've learnt method overloading, which is termed the static binding, that is, static polymorphism, because which method is invoked can be resolved at compiling time, not run time. Two differences between dynamic and static polymorphisms are as below:

Static:
(1) Method Overloading, within a class, several methods share a name.
(2) Which method is invoked is resolved at compilation time.

Dynamic:
(1) Method Overriding, a given method in the base class is redesigned and kept its method head unchanged in its derived class.
(2) Which method is invoked is resolved at run time.

哪个 treatPatient() 方法会被调用呢？调用的是 Surgeon 的 treatPatient()，因为引用指向的是 Surgeon 的对象。当基类的引用被用来调用方法时，到底哪个方法被调用，是根据引用所指向的对象，而不是引用本身决定的。

尽管 surgdoc 是 Doctor 类的引用，但调用的是 Surgeon 的方法，因为它指向的是 Surgeon 的对象，这是在运行时确定的，因此称之为动态多态或者运行时绑定。动态多态是通过更为概括性的引用（基类的引用）与一个具体的对象的关联实现的。在第 3 章，我们学习了方法的重载，那是静态绑定，因为具体哪个方法被调用是在编译时，而不是在运行时决定的。静态与动态多态的两个不同之处：

静态：

（1）方法重载，在一个类内，多个方法名字相同。

（2）其中哪一个方法被调用是在编译时确定的。

动态：

（1）方法重写（覆盖），子类的方法和基类的方法使用一个名字，方法头部保持不变。

（2）哪个方法会被调用是在代码运行时确定的。

```
Doctor obj=new Surgeon();
obj.treatPatient(); //Surgeon's method is invoked
```

【Example 5-17】Overloading vs. overriding

```
package org;
import kkk.*;
public class Shape {              //overloading
  public int getSides() {
    return 0;
  }
  public int getSides(int i) { //overloading
    return i;
  }
  public static void main(String[] args) {
    Shape shape;
    shape=new Triangle();
    System.out.println("My shape has "+shape.getSides()+"sides.");
    shape=new Rectangle();
    System.out.println("My shape has"+shape.getSides(4)+"sides.");
  }
}

package kkk;
import org.*;
public class Triangle extends Shape {
  public int getSides() {         //overriding
    return 3;
  }
}
package kkk;
import org.*;
public class Rectangle extends Shape {
  public int getSides(int i) { //overriding
    return i;
  }
}
```

Output:

My shape has 3 sides.

My shape has 4 sides.

【Example 5-18】**Polymorphism application**

```
public class Music {
  public static void tune(Instrument i) {
    i.play('C');
  }
  public static void main(String[] args) {
    Wind flute=new Wind();
    tune(flute);
    Instrument piano=new Instrument();
    tune(piano);
  }
}
class Instrument {
  public void play(char n) { //overridden in Wind
    System.out.println("Instrument.play() "+n);
  }
}
class Wind extends Instrument {
  public void play(char n) { //different implement, same signature
    System.out.println("Wind.play() "+n);
  }
}
```

Output:

Wind.play() C

Instrument.play() C

Where, Wind is an Instrument. The method Music.tune() takes an argument of the Instrument reference. The actual parameter can be any object of class derived from Instrument.

【Example 5-19】**Dynamic binding**

```
class Shape {
  public void draw() {}
  public void erase() {}
}
class Circle extends Shape {
  public void draw() {System.out.println("Circle.draw()");}
  public void erase() {System.out.println("Circle.erase()");}
}
class Square extends Shape {
  public void draw() {System.out.println("Square.draw()");}
  public void erase() {System.out.println("Square.erase()");}
}
class Triangle extends Shape {
```

```
    public void draw() {System.out.println("Triangle.draw()");}
    public void erase() {System.out.println("Triangle.erase()");}
}
public class Shapes {
  public static void main(String[] args) {
    Shape[] s={new Circle(),new Square(),new Triangle()};
    for(Shape k:s){k.draw();k.erase();}
  }
}
```

【Example 5-20】**Override method**

```
public class Override {
  public static void main(String[] args) {
    Sub sb=new Sub();
  }
}
class Super {
  public Super() {
    System.out.println("Super constructor");
    test(); //which test is called?
  }
  void test() {
    System.out.println("test() in super");
  }
}
class Sub extends Super {
  public Sub() {
    System.out.println("Sub constructor");
  }
  void test() {
    System.out.println("test() in sub");
  }
}
```

Output:

Super constructor

test() in sub

Sub constructor

Change "Sub sb=new Sub();" to "Super sb=new Sub();" in main, the output keeps unchanged.

5.6.3 Referring to a member of the super class by super keyword（用 super 指向基类成员）

The keyword super can be used in three ways:

(1) Respond to a base class' constructor (super();), the "super();" must be the first statement in a constructor.

(2) Invoke a base class' method (super.), the "super." can be anywhere.

(3) Refer to a base class' field (super.), the "super." can be anywhere.

在 3 种情况下使用 super 关键字：

（1）代表基类的构造方法，"super();"必须是构造方法的第一条语句。

（2）用于调用基类的方法，"super."可以处于任意位置。

（3）用于指向基类的 field "super."可以处于任意位置。

If the treatPatient() in the Surgeon class wants to do the functionality defined in Doctor class and then perform its own specific functionality, the treatPatient() method in the Surgeon class could be written as:

```
void treatPatient() {
  super.treatPatient();
                //add codes specific to Surgeon
}
```

【Example 5-21】**Calling method of super class**

```
class Base {
  void p(double i) {
    System.out.print(i*2+" ");
  }
}
class A extends Base {
  void p(double i) {
    super.p(i);
    System.out.println(i);
  }
}
public class Test {
  public static void main(String args[]) {
    A a=new A();
    a.p(1);
    Base b=new A(); //runtime binding
    b.p(2);
    Base c=new Base();
    c.p(3);
  }
}
```

Output:

2.0 1.0

4.0 2.0

6.0

【Example 5-22】**Calling super constructor and method**

```
public class Base {
  String name;
```

```
    void bfun() {
      System.out.println("name="+name);
    }
    Base(String n) {
      name=n;
    }
    public static void main(String[] args) {
      SSA k=new SSA("Wang",102);
      k.bfun();
    }
}
class SSA extends Base {
    private int number;
    SSA(String n,int numb) {
      super(n);
      number=numb;
    }
    void bfun() {
      super.bfun();  //override bfun()
      System.out.println("name="+name);
      System.out.println("number="+number);
    }
}
```

5.6.4 Hiding fields and static methods of the base class
（隐藏静态方法和 fields 的基类）

Fields cannot be overridden, but they can be hidden, for instance. if you declare a field in a subclass with the same name as one in the super class, the super class' field can only be accessed by the keyword super in the derived class. Polymorphism only relates methods.

  fields 不能被重写，但可以被隐藏。如果在子类里重新声明了基类里的一个 field，则其基类的 field 被隐藏，基类的被隐藏的 fields 在派生类里只能通过关键字 super 访问。多态仅仅针对方法。

【Example 5-23】Use "super."to invoke the base class' fields and methods

```
public class Test {
    int a=1,b=2,c=3;
    void fun(){
      a=2*a;
      b=2*b;
      c=2*c;
    }
    public static void main(String args[]) {
      TT a=new TT();
      a.out();
    }
}
class TT extends Test {
    int a=11,b=12,c=13;
```

```
  void fun(){
    super.fun();
    a++;
    b++;
    c++;
  }
  void out(){
    fun();
    System.out.print("a="+a+" b="+b+" c="+c+'\n');
    System.out.print("a="+super.a+" b="+super.b+" c="+super.c+'\n');
    //access super's fields
  }
}
```

Output:

a=12 b=13 c=14

a=2 b=4 c=6

The final methods cannot be overridden.

An instance method can be overridden only if it is accessible. Thus a private method cannot be overridden, because it is not accessible outside its own class. If a method defined in a subclass is private in its super class, the two methods are completely unrelated.

Like an instance method, a static method can be inherited. However, a static method cannot be overridden. If a static method defined in the super class is redefined in a subclass, the method defined in the super class is hidden.

不允许重写 final 方法。一个实例方法之所以能够被重写是因为它可以在类外被访问。私有方法不能在类外访问，因此不能被重写。如果在子类定义了一个与基类相同的私有方法，则这两个方法没有关系。

与实例方法相同，静态方法也被子类继承，但它不能被重写。静态方法也可以被隐藏，如果在子类重新定义一个与基类相同的静态方法，则基类的静态方法被隐藏，基类和子类分别使用自己的静态方法。

【Example 5-24】**Static method cannot be overridden, but be hidden**

```
class Sup {
  private int a=100;
  int base=2;
  void deft() { System.out.println("I am a default method.");}
  int getA(){return a;}
  public static void aaa() {
    System.out.println("I am Super.");
  }
}
class Subb extends Sup {
  private int a=20;
  int base=8;
```

```
    int getA() {
      return a;
    }
    public static void aaa() {
      System.out.println("I am Sub");
    }
  }
  public class Test {
    public static void main(String args[]) {
      Sup.aaa();
      Sup a=new Sup();
      System.out.println("value a="+a.getA());
      Subb.aaa();
      Sup b=new Subb();
      System.out.println(b.base);
      b.deft();
      System.out.println("value a="+b.getA());
    }
  }
```

Output:

I am Super.
value a=100
I am Sub
2
I am a default method.
value a=20

Exercises

1. Write a program to print a graduate student's information.

(1) Define a Student class and a Graduate class that is derived from Student and is the main class.

(2) Student has two fields: student's Id and name, which are initialized through Student's constructor.

(3) Override toString method, let print(an object) produce an output like "学号 :101 姓名 :David Bissell".

(4) Graduate has one field: supervisor that is initialized through Graduate's constructor.

(5) Method main can be:
```
public static void main(String[] args) {
  Graduate stu=new Graduate(101,"David Bissell","Jeff Trump");
  System.out.print(stu+" 导师 :"+stu.supervisor);
}
```

写程序，输出一个研究生的信息。

(1) 定义 Student 类和 Graduate 类，Graduate 是主类，继承 Student 类。

(2) Student 类有两个 fields：学号和姓名，在 Student 的构造方法里初始化它们。

(3) 重写 toString，当 print(对象) 时，输出类似 "学号：101 姓名：David Bissell"。

(4) Graduate 有一个 field：supervisor，在 Graduate 的构造方法里初始化它。

(5) 主方法如上。

2. Understand the sequence of constructors' calling.

(1) Define classes: Comp and Stem. Each class has a field and a constructor to initialize it and print it. Each constructor increases the field "sequence" in Root by 1, and prints it.

(2) Root is the main class, Root is as below.

(3) Stem is a subclass of Root. Stem also contains an object of Comp.

```
class Root {
  String rs;
  static int sequnce=0;
  Comp comp=new Comp("Comp in Root");
  Root(String str) {
    sequnce++;
    rs=str;
    System.out.println("I'm Root, sequence="+sequence);
    System.out.println(rs);
  }
  public static void main(String[] ss) {
    Stem stem=new Stem("to Root", "to Stem");
  }
}
```

理解构造方法的调用次序。

(1) 定义 2 个类：Comp 和 Stem。每个类都有一个 field，有一个对这个 field 初始化并输出的构造方法，构造方法还要将 Root 类的一个静态变量 "sequence" 加 1 并输出。

(2) Root 是主类，Root 的定义见上面的代码。

(3) Stem 是 Root 的子类，它也有一个 Comp 类的 field。

3. Understand the final keyword.

(1) Correct following codes and tell the errors' reasons.

(2) Why the two statements with question marks are right?

理解 final 关键字。

(1) 改正下面的程序，说明错误原因。

(2) 说明为什么带问号的两条语句是正确的？

```
class Value {
  int i;
  public Value(int k) {
    i=k;
  }
```

```java
}
public class FinalData {
  public FinalData(int ik) {
    V_ONE=ik;
    V2=new Value(22);
    V3=new Value(33);
    System.out.println(ik);
  }
  private final int V_ONE=12;
  private static final int V_TWO;
  public static final int V_THREE=39;
  private final Value V2;
  private final Value V3=new Value(33);
  private final int[] a={1,2,3,4,5,6};
  public static void main(String[] args) {
    FinalData fd1=new FinalData(78);
    fd1.V2.i++; //?
    for(int i=0; i<fd1.a.length; i++) {
      fd1.a[i]++; //?
      System.out.print(fd1.a[i]+" ");
    }
    System.out.println();
    System.out.println(V_TWO);
    FinalData fd2=new FinalData(29);
    System.out.println(V_THREE);
    System.out.println(fd2.V_ONE);
  }
}
```

4. Whether the statements with question marks are right? If not, correct them.

判断带"？"的语句是否正确，如果不正确，请修改。

```java
class Foo {
  private boolean x;
  boolean y=true;
  public void xyz() {
    System.out.println(x); //?
    System.out.println(convert()); //?
  }
  private int convert(){return x? 1: -1;}
}
public class TestA {
  public void abc() {
    Foo foo=new Foo();
    System.out.println(foo.x); //? If not, delete this line.
    System.out.println(y); //?
    System.out.println(convert()); //?
  }
  public static void main(String[] a) {
    Foo foo=new Foo();
    xyz(); //? If not, delete this line.
```

```
        Foo.xyz(); //?
        abc(); //? If not, create an object and then correct.
        System.out.println(foo.y); //?
    }
}
```

5. Understand "protected".

(1) Define a class named Father that has a protected field, define another class in the same file with a method that uses the protected field of Father (without the main method in both of the two classes).

(2) In another package define a method to print the protected field of Father. In main, create an object of class Son to call the method.

```
finish the codes below.
class Son extends Father {
    …
    Public static void main(String ss) {
        …
    }
}
```

理解访问修饰符 protected。

(1) 定义 Father 类，类里有一个 protected 成员，再定义一个类，类中使用 Father 的保护的成员（2个类都不含 main 方法）。

(2) 在另一个包里定义 Son 类，在其中定义一个方法，输出 Father 的 protected 成员。在 main 里创建 Son 的对象并调用这个方法。

6. Add a class, named A to the following codes. Class A is a subclass of Person. In A, define a field (age), a constructor and a method which overrides the pfun method of Person, use super key in the method, output the three fields.

在下面的代码中，增加一个名字为 A 的类。A 是 Person 的子类。在 A 中定义一个 field（age），一个构造方法，并重写 Person 里的 pfun 方法，在方法中使用 super 关键字，输出 3 个 fields。

```
public class Person {
    private int id;
    private String name;
    void pfun(){
        System.out.println("id="+id);
        System.out.println("name="+name);
    }
    Person(int a, String b) {
        id=a;
        name=b;
    }
    public static void main(String[] args) {
        A k=new A(102,"Wang",22);
        k.pfun();
    }
}
```

Chapter 6
Abstract Class and Interface(抽象类和接口)

Introducing abstract classes and interfaces allow you to define a code of behaviors that can be conducted by any classes, and then manipulate them by unified commands.

For certain type of tasks, an interface tells what have to do by specifying the signatures (names, argument lists) of methods. For instance, dealing with batch of data, the data can be quite different at types, such as data for students, doctors, soldiers and so on, but all they have common behaviors such as adding to a collection, sorting, searching and so on. As to the details how to do are the method bodies' business and the classes that implement the interface must design methods of the interface.

Abstract classes and interfaces are useful in three aspects.

(1) Capturing similarities among unrelated classes without artificially forcing a class relationship.

(2) Declaring methods that one or more classes are expected to implement.

(3) A class can implement multiple inheritances, which is a feature of some object-oriented languages owning.

引入抽象类和接口，允许定义统一的行为规范，以方便用相同的方式进行操控，具体的行为则由任意类实施。

对于某些特定类型的任务，接口通过规定方法的基本特征（名字，参数列表）来指定必须要做的事情。例如，批量处理学生的、医生的、士兵等完全不同类型的数据，其基本的处理方法，如输入、排序、查找等却是一样的。至于如何去做这些事情，是方法体的事，实现接口的类，必须设计接口的所有方法。

抽象类和接口的用处体现在以下 3 个方面。

（1）提取彼此不相关的类的共同特性。

（2）声明一个或多个类要实现的方法。
（3）一个类可以实现多个接口，与一些面向对象语言所拥有的多重继承的特性一致。

6.1 Abstract class（抽象类）

6.1.1 Abstract method（抽象方法）

If a method defined in a class has only the method's declaration (method head) but has no the method body, is called an abstract method, the keyword "abstract" must be put in front of the method. An abstract method cannot be modified with private or static keyword. An abstract method is defined as below:

```
abstract methodtype methodname(argument list);
```

定义类的方法时，如果只有方法头而无方法体，该方法被称为抽象方法，必须在方法前面加关键字 abstract。抽象方法不能用 private 或 static 修饰。定义抽象方法的格式如下：

```
abstract 方法类型 方法名（形参列表）;
```

Abstract methods are typically used in abstract classes or interfaces. Abstract methods stay in abstract classes or interfaces.

【Example 6-1】**Defining an abstract method in a class**

```
abstract class A {
  public abstract void result();
}
```

6.1.2 Abstract class（抽象类）

Abstract classes are modified with the keyword "abstract", the syntax is as below:

```
abstract class ClassName {
  class body
}
```

The abstract classes can be divided into two forms. If a class contains one or more abstract methods, the class must be defined as an abstract class. Even if a class does not contain any abstract methods, the class can be defined as an abstract class if needed. Example 6-2 and 6-3 respectively represent the two types of abstract classes.

抽象类可以分为两种。如果类里包含抽象方法，该类必须被定义为抽象类。不包含抽象方法的类，根据需要也可以把它定义为抽象类。例 6-2 和 6-3 分别对应这两种抽象类。

【Example 6-2】**Define an abstract class that has an abstract method**

```
abstract class Flower {
  public abstract void result();
}
```

【**Example 6-3**】 **Define a class without any abstract methods**

```
abstract class Flowers {
  public String fName;
    public Flowers(String n) {
    fName=n;
  }
  public void result() {
    System.out.println(fName+" Yellow Flower.");
  }
}
```

The only usage of an abstract class is as the super class of the other classes, so never try to create an object of an abstract class. The subclass of an abstract class should implement all the abstract methods of its super abstract class, that is, defining all the declared methords normally, or else, the subclass is still an abstract one. Of course, as a super class, the reference of a super abstract class can reference to its subclass' objects. Example 6-4 is an example of defining a subclass which inheritances the class defined in example 6-2 and implements the abstract method.

抽象类的唯一用途是做其他类的父类，因此不要试图用抽象类创建对象。抽象类的子类应实现其抽象父类的所有抽象方法，即在子类定义所有声明了的方法，否则，该子类还是抽象类。当然，作为父类，抽象类的引用可以指向其子类的对象。例6-4定义了一个类，它继承例6-2里的抽象类并实现了它的抽象方法。

【**Example 6-4**】 **Create a subclass that implements the abstract method in Example 6-2**

```
class SubFlower extends Flower {
  public void result() {
    System.out.println("SubFlower implements abstract result().");
  }
}
public class Sam6 {
  public static void main(String[] args) {
    Flower a=new SubFlower();
    a.result();
  }
}
```

Output:

SubFlower implements abstract result().

6.2 Interface（接口）

6.2.1 Introduction（简介）

In Java, a class at most can have one direct super class. However, in practical applications, multiple inheriting is often needed. Therefore, Java allows a class directly implements multiple interfaces to realize the multiple inheriting indirectly.

Java 是单继承，即一个子类最多只能有一个直接父类。但是在实际应用中，常常需要多继承来解决问题，因此，Java 允许一个类直接实现多个接口，间接地实现了多继承。

6.2.2 Defining interfaces（接口的定义）

The purpose of an interface is to define a code of behaviors that can be conducted by any classes, that is, to unify what to do.

The keyword "interface" is used to define an interface. The definition of an interface is similar to the definition of a class. It is also composed of interface declaration and interface body. The syntax is as below:

```
[modifier] interface interfaceName {
  interface body
}
```

Where, modifier is optional and can be public or default (no modifier). If public, the defined interface is public and available anywhere. If default, the interface can only be accessed in the package where it is defined.

其中，modifier 是可选的，可以是 public 或 default（无修饰）。如果是 public，则该接口在任何地方都可用，如果是 default，则该接口只能在定义接口的包中被访问。

An interface body contains constans and methods. The constants are public static and final by default. If there are any constants in an interface, the constants must be assigned values at the defining time. Methods in an interface are public and abstract by default and can be omitted. Example 6-5 is an interface example.

接口体中包含常量与方法。常量在默认情况下是 public static final 的。如果接口包含常量，必须在定义常量时给它们赋值。接口中的方法默认是 public abstract 的，可以省略 public 和 abstract。例 6-5 是一个接口的例子。

【Example 6-5】Define an interface named Circle, it contains a constant PI and two methods.

```
public interface Circle {
  double PI=3.1415926;
  double perimeter(double r);
  double area(double r);
}
```

The codes are equivalent to the codes below.

```
public interface Circle {
  public static final double PI=3.1415926;
  public abstract double perimeter(double r);
  public abstract double area(double r);
}
```

A constant defined in an interface can be accessed by the interface name, for instance, Circle.PI. The source file name is Circle.java and the bytecode file name is Circle.class, which follows the same rules as the class.

6.2.3 Implementation of interfaces（接口的实现）

We can only use an interface to define a reference, not create an object, because the methods in the interface are all abstract. Defining a class that implements an interface must design all the methods of the interface, if not, the class must be declared abstract. A class can implement one or more interfaces at the same time. If multiple interfaces are implemented, the interface names are separated by commas ",". The syntax of defining a class that implements multiple interfaces is as below.

```
class ClassName implements InterfaceName1, InterfaceName2, …{
  //implement all methods in interfaces
}
```

只能用接口定义引用，不能创建对象，因为接口的方法都是抽象的。定义一个实现接口的类时必须实现接口的所有抽象方法，否则，要将该类声明为抽象类。一个类可以同时实现一个或多个接口。如果实现多个接口，接口名之间需要用逗号","隔开。

Example 6-6 defines a class that implements the interface defined in example 6-5, example 6-7 defines a class that implements the interface defined in example 6-5 too, but it failed to define the abstract method area() of the interface, it can only be declared as the abstract class.

例 6-6 定义了一个类实现了例 6-5 中定义的接口，例 6-7 也定义了一个类实现例 6-5 中定义的接口，但由于没有实现接口的抽象方法 area()，必须将该类声明为抽象类。

【Example 6-6】 **Define a class that implements the interface defined in example 6-5**

```
class Circle1 implements Circle {
  public double perimeter(double r) {
    return 2*PI*r;
  }
  public double area(double r) {
    return PI*r*r;
  }
  public static void main(String[] args) {
    Circle a=new Circle1(); //upcast
    System.out.println("Circle a perimeter:"+a.perimeter(2.5));
    System.out.println("Circle a area:"+a.area(2.5));
    Circle1 b=new Circle1();
    System.out.println("Circle1 b perimeter:"+b.perimeter(2.1));
    System.out.println("Circle1 b area:"+b.area(2.1));
  }
}
```

Output:

Circle a perimeter:17.2963
Circle a area:21.620375
Circle1 b perimeter:14.528892
Circle1 b area:15.255336600000001

【**Example 6-7**】 **An abstract class failed to implement the abstract area() method**

```
abstract class Circle2 implements Circle {
  public double perimeter(double r) {
    return 2*PI*r;
  }
}
```

In a class that implements an interface, all the abstract methods of the interface must be defined and keep all the signatures, such as method names, parameter lists and return types of the methods exactly the same as that in the interface. Example 6-8 cannot be compiled successfully, because the method perimeter() takes different parameter list from the one in interface Circle and method area() is public in interface Circle but in class Circle1, it is default.

类的实现接口的一个类中，接口中的所有抽象方法必须被定义，并且方法的特性如方法名、参数列表和返回类型要与接口中的抽象方法完全一致。例 6-8 不能成功通过编译，是因为方法 perimeter() 与接口 Circle 里的方法参数列表不同，方法 area() 在接口中是 public 的，而在 Circle1 中是 default。

【**Example 6-8**】 **The methods perimeter() and area() cannot be implemented successfully**

```
class Circle1 implements Circle {
  double r;
  public Circle1(double r) {
    this.r=r;
  }
  public double perimeter() {    //takes different parameter list
    return 2*PI*r;
  }
  double area(double r) {        //public in interface Circle
    return PI*r*r;
  }
}
```

6.2.4 Comparation of interfaces and abstract classes
（接口与抽象类的比较）

Except for containing abstract methods, an abstract class is the same as a normal class, of course can't create any objects with it. An interface contains only abstract methods and constants. Therefore, an interface is a special abstract class. The specific comparison is as shown in Table 6-1.

抽象类除了包含抽象方法以外,与普通类没有区别,当然不能用它创建对象。接口仅包含抽象方法和常量,因此,接口是一种特殊的抽象类。它们的异同如表 6-1 所示。

Table 6-1   Similarities and differences between interfaces and abstract classes

	Abstract class	Interface
grammar	abstract class ClassName{}	interface InterfaceName{}
method	contain both non-abstract and abstract methods or only non-abstract methods	contain abstract methods only
static method	can contain	no
field	can contain variables and contants	only contants
constructor	can contain	no
member access	may or may not public	public

6.3 Inner class（内部类）

Inner classes can make programs simple and concise.

Java allows nested class definition, that is, you can define a class inside another class, and the inside class is called the inner class. The inner class acts as a member of the outside class, and the outside class is called the outer class.

A non-inner class can only be public and default. However an inner class can be declared public, default, protected or private subject to the same visibility rules applied to a member of the class.

内部类可以使程序变得简单和简洁。

Java 类的定义允许嵌套,即可以在一个类内定义类,这个在类内部被定义的类称为内部类。内部类被看成是外部类的成员。外层的类称为外部类。

非内部类只能是 public 或者 default,然而内部类可以是 public、default、protected 或者 private,与普通的类成员一致。

6.3.1 Members in inner class（内部类成员）

【Example 6-9】Define a inner class

```java
class OutClass {
  String str1="I'm outside.";
  InClass ins=new InClass();
  private String getStr1() {
    return str1;
  }
  class InClass {
    String str2="I'm inside.";
    void getStr2() {
```

```
      System.out.println(getStr1()+" "+str2);
    }
  }
}
public class Test {
  public static void main(String[] args) {
    OutClass ous=new OutClass();
    ous.ins.getStr2();
    OutClass.InClass in=ous.new InClass(); //Create an inner class object "in"
    in.getStr2();
  }
}
```

Output:

I'm outside. I'm inside.

I'm outside. I'm inside.

Example 6-9 is an inner class example. Under the "src" folder, only one source file Test.java there, but under the "bin" folder, three bytecode files are produced. Pay attention to the inner class's name Outclass$InClass.class.

OutClass$InClass.class

OutClass.class

Test.class

例 6-9 是个内部类的例子，src 文件夹下的一个源文件 Test.java 对应 bin 文件夹下的 3 个 bytecode 文件，请特别注意内部类的名字。

If an inner class has a field that takes the same name as the field of the outer class, you can access to the outer field in the inside class by format: OutClassName.this.fieldname. Example 6-10 shows how to access the same name fields.

如果内部类与外部类中有同名的 field，可以在内部类里通过"外部类名.this.成员名"来访问同名的外部类中的 field。例 6-10 是访问同名 field 的例子。

【Example 6-10】Access to the fields of the inner and outer classes

```
class OutClass {
  String str="outer field";
  class InClass {
    String str="inner field";
    void g(String str) {
      System.out.println(str);
      System.out.println(this.str);              //field of inner class
      System.out.println(OutClass.this.str);  //outer field
    }
  }
}
public class T3 {
  public static void main(String[] args) {
    OutClass ous=new OutClass();
```

```
        OutClass.InClass in=ous.new InClass();
        in.g("Hello!");
    }
}
```

Output:

Hello!
inner field
outer field

An inner class can access the fields and methods defined in the outer class in which it nests, so you need not to pass the reference of the outer class to the constructor of the inner class. Example 6-11 is such an example.

内部类可以访问其外层类的 fields 和方法，因而无须将外层类的引用传给内部类的构造方法。例 6-11 是内部类访问外部类的 field 和方法的例子。

【Example 6-11】**Access to outer members**

```
public class Outer {
  private int data;
  public void m() {
    System.out.println("We are outer method and data"+data);
  }
  public class Inner {
    public void me() {
       data++;
       m();
    }
  }
  public static void main(String[] args) {
    Outer ot=new Outer();
    Outer.Inner iner=ot.new Inner();
    iner.me();
  }
}
```

Output:

We are outer method and data 1

6.3.2 Local inner class（局部内部类）

A local inner class is equivalent to a local variable of a method.

【Example 6-12】**External class' object refers to a local class' method**

```
class OutClass {
  String str=" 外部类成员 ";
  void ff() {
    class InClass {
```

```
      void dd() {
        System.out.println("method of local inner class: dd()");
      }
    }
    InClass x=new InClass();
    x.dd();
  }
}
public class T3 {
  public static void main(String[] args) {
    OutClass ous=new OutClass();
    ous.ff();
  }
}
```

Output:

method of local inner class: dd()

6.3.3　Anonymous inner class（匿名内部类）

An anonymous inner class is a special inner class. It has no class name and an object of the class is created at the time of the class definition. Usually they are for class inheritance and interface implementing. Example 6-13, 6-14 realize class inheritance with the local inner class and the anonymous inner class respectively. Example 6-15, 6-16 implement interface with the local inner class and the anonymous inner class respectively.

匿名内部类，是一种特殊的内部类，它没有类名，在定义的同时就创建了该类的一个对象，通常用于类的继承或接口的实现。例 6-13 和 6-14 分别用局部内部类和匿名内部类实现了类的继承。例 6-15 和 6-16 分别用局部内部类和匿名内部类实现了接口。

【Example 6-13】**Local inner class for inheritance**

```
class T2 {
  void w() {
    System.out.println("w() in T2 class.");
  }
}
public class T3 {
  public static void main(String[] args) {
    class InClass extends T2 { //local inner class
      void w() {
        System.out.println("w() in InClass.");
      }
    }
    InClass y=new InClass();
    y.w();
  }
}
```

Output:

w() in InClass.

【Example 6-14】 **Anonymous class for inheritance**

```
class T2 {
  void w() {
    System.out.println("w() in T2 class.");
  }
}
public class T3 {
  public static void main(String[] args) {
    T2 y=new T2() {
      void w() {
        System.out.println("w() in InClass.");
      }
    };
    y.w();
  }
}
```

Output:

w() in InClass.

【Example 6-15】 **Local inner class implements an interface**

```
interface T2 {
  void w();
}
public class T3 {
  public static void main(String[] args) {
    class InClass implements T2 { //local inner class
      public void w() {
        System.out.println("w() in InClass.");
      }
    };
    InClass y=new InClass();
    y.w();
  }
}
```

Output:

w() in InClass.

【Example 6-16】 **Anonymous class implements an interface**

```
interface T2 {
  void w();
}
public class T3 {
```

```
    public static void main(String[] args) {
      T2 y=new T2() { //anonymous inner class
        public void w() {
        System.out.println("w() in InClass.");
        }
      };
      y.w();
    }
}
```

Output:

w() in InClass

Exercises

1. Write a program to understand the abstract class.

(1) Define an abstract class Father, it has an abstract protected void print() method.

(2) Derive a class Son that implements the method print().

(3) Define a class Test, it has a static method "test" that takes a reference argument to the base class Father, inside method test, call print().

(4) Class Test is the main class. In method main, create an object of Son, and call method test.

写程序，理解抽象类。

(1) 定义一个 Father 类，它有个 abstract protected void print() 方法。

(2) 派生一个 Son 类，实现 print() 方法。

(3) 定义一个 Test 类，Test 是主类，其中包含 static 方法 test，方法有个指向基类 Father 的引用参数，在方法中调用 print()。

(4) Test 是主类，在 main 中创建一个 Son 对象，并通过它调用 test 方法。

2. Write a program to understand the interface.

(1) Define an interface, in which a method void point() is declared.

(2) Define a main class, which has 2 fields, and implements the point() of the interface. The point() prints the 2 fields of an object, for example "point: 12.5, 23.0".

(3) In main, create an object and by which the point() is invoked.

写程序，理解接口。

(1) 定义一个接口，在其中声明 void point() 方法。

(2) 定义主类，主类里包含 2 个 fields，并实现 point() 方法。point() 输出一个对象的 2 个 fields，例如，"point: 12.5, 23.0"。

(3) 在 main 里创建一个对象，并通过它调用 point() 方法。

3. Read, complete and run the program.

```java
interface History {
  void u();
}
interface Exam {
  void v();
}
interface Labs {
  void w();
}
interface Work extends History, Exam, Labs {
  void x();
}
abstract class Doctor {
  abstract void doc();
}
class Test extends___(1)___implements___(2)_____ {
  public void u() { System.out.println("I am u()"); }
  public void v() { System.out.println("I am v()"); }
  public void w() { System.out.println("I am w()"); }
  public void x() { System.out.println("I am x()"); }
  public void doc() { System.out.println("I am doc()"); }
}
public class ch6 {
  public static void m1(History h) { h.u(); }
  public static void m2(Exam exam) { exam.v(); }
  public static void m3(Labs labs) { labs.w(); }
  public static void m4(Work work) { work.x(); }
  public static void m5(Doctor d) { d.doc(); }
  public static void main(String[] args) {
    ___(3)___ s=new___(4)___();
    m1(s);
    m2(s);
    m3(s);
    m4(s);
    m5(s);
  }
}
```

Chapter 7 Generics and Collections(泛型与集合)

7.1　Generics(泛型)

The generic type can be used for classes, interfaces and methods. It is parameterized over types. In Java library, there are a group of container classes using generic parameters for dealing with objects.

泛型可用于类、接口和方法，它将类型参数化。在 Java 库中，有一组容器类使用泛型参数来处理对象。

7.1.1　Concept of generics(泛型的概念)

Declaring variables or defining methods, you have to specify the concrete types of the variables or the argument list of the methods, like:

```
class A {
  Integer i;
  String s;
  void fun(String a, Integer b) {…}
}
class B {
  A i;
  String s;
  void fun(String a, A b) {…}
}
```

Ordinary classes and methods work with specific types: either primitive or reference types. At times, different types of data need to be dealt by the same groups of methods, like sorting a group of data. The type rigidity constrains the idea to use the same method to deal with different types of data. Data type generalizations can be realized by the method overloading. However, that means you write many codes alike for different data types. If the type parameter, that is "some unspecified types", is introduced, you can write codes that might be used across more types. That is the concept of generics.

普通的类与方法处理特定类型的数据：基本数据类型或引用类型。有时需要用一组方法处理不同的数据，例如对一组数据排序。对类型的严格限制，使得用相同方法处理不同类型数据的思想难以实现。可以通过方法的重载实现数据类型的泛化，然而这意味着针对不同的数据类型写许多雷同的代码。如果引入类型参数，即"未指定的类型"，就可以写出能够处理非特定数据类型的代码，这就是泛型。

7.1.2　Generic classes（泛型类）

The type parameter is not specified to a specific data type. Defining a generic class, the type parameter must be inside the angle brackets after the class name. The syntax defining a generic class is as below.

```
class A<T> {…} //where T is the type parameter
or
class B<T1,T2> {…}
class C<X1, X2> {X1 a; X2 b;} //X1,X2 are type parameters
Where A, B and C are the generic classes.
```

The purpose of defining classes is to create the objects. Creating objects with the generic classes, you must explicitly tell the actual data types you'll deal with.

For instance:

```
C<String, Integer> m=new C<String, Integer>();
```

Where, type parameters X1 and X2 are substituted by String and Integer.

类型参数不代表特定类型。定义泛型类时，要把类型参数放到类名后面的三角括号里。定义类的目的是创建对象，在使用泛型类创建对象时必须用实际类型替换类型参数。例 7-1 是泛型类的例子。

【Example 7-1】Generic class

```
class Atest {
  int a,b;
  public Atest(int x,int y) {a=x; b=y;}
}
public class Holder<T> {
  private T a;
  public void set(T k) {a=k;}
  public T get() {return a;}
  public static void main(String[] args) {
    Holder<Atest> h=new Holder<Atest>();
    Holder<String> m=new Holder<String>();
    Holder<Integer> n=new Holder<Integer>();
    h.set(new Atest(10,20));
    Atest at=h.get();
```

```
        System.out.println("I'm Atest:"+at.a+" and "+at.b);
        m.set("The string");
        System.out.println("I'm String:"+m.get());
        n.set(12); //up boxing
        System.out.println("I'm Integer:"+n.get());
    }
}
```

 Output:

I'm Atest:10 and 20
I'm String:The string
I'm Integer:12

In example 7-1, with the same two methods "set" and "get" of class Holder, three totally different types of data are dealt with. You must specify the type you want to use while creating an object. The core idea of Java generics: You tell what data type you want to use, and it takes care of the details such as data type checking. If you refuse to tell the data type while creating an object, without the angle brackets after the class name, the actual types are Object by default.

在例 7-1 中，Holder 类的 2 个方法 set 和 get 处理了 3 种完全不同的数据。在创建对象时必须指定数据类型。泛型的核心思想是：用户指定类型，其他细节如类型检查由泛型机制处理。如果在创建对象时未能指定数据类型，即在类名后面没有三角括号，则实际类型默认为 Object。

In general, you can treat generics as if they are any other type — they just happen to have type parameters. You can use generics just by naming them along with their type argument list. For instance:

```
public class Two<A,B> {
  public final A first; //first is an A type constant
  public final B second;
  public Two (A a,B b) {first=a; second=b;}
  public String toString() {
    return "("+first+", "+second+")";
  }
}
```

You can add more type parameters in a sub generic class, for instance:

```
public class Three<A,B,C> extends Two<A,B> {
 public final C third;
  public Three (A a,B b,C c) {
    super(a,b);
    third=c;
  }
  public String toString() {
    return "("+first+", "+second+", "+third +")";
  }
}
```

【Example 7-2】 **Define generic class and its subclass**

```
class Amp {}
public class TwoTest {
  static Two<String, Integer>f() {
    return new Two<String, Integer>("hi", 47); //converts the int to Integer
  }
  static Three<Amp, String, Integer>g() {
    return new Three<Amp, String, Integer>(new Amp(),"Hello",3);
  }
  public static void main(String[] args) {
    Two<String,Integer> ttsi=f();
    System.out.println(ttsi);
    // ttsi.first="there"; //Compile error
    System.out.println(g());
  }
}
class Two<A,B> {
  public final A first;
  public final B second;
  public Two (A a, B b) {first=a; second=b;}
  public String toString() {
    return "(" + first + ", " + second + ")";
  }
}
class Three<A,B,C> extends Two<A,B> {
  public final C third;
  public Three (A a, B b, C c) {
    super(a, b);
    third=c;
  }
  public String toString() {
    return "("+first+", "+second+ ", "+third+")";
  }
}
```

 Output:

(hi, 47)

(Amp@6bade9, Hello, 3)

【Example 7-3】 **Without actual types while creating an object, Eclipse gives warning information**

```
public class Test<T,E> {
  private T a;
  private E b;
  public Test(T k, E g) {
    a=k; b=g;
  }
  public T getT() { return a;}
```

```
    public static void main(String[] args) {
    Test<String, Integer> s=new Test<String, Integer>("I'm type T",
new Integer(12));
    System.out.println(s.getT());
    Test<Test1,Double> d=new Test<Test1,Double>(new Test1(),
new Double(23.8));
    System.out.println(d.getT());
    Test t=new Test(new Test1(), new String("It's me!")); //Actual types are Object
      System.out.println(t.getT());
    }
}
class Test1 {
    public String toString() {
      return "I'm Test1.";
    }
}
```

Output:

I'm type T
I'm Test1.
I'm Test1.

7.1.3　Type parameters use "extends" and "super" keywords
（类型参数中使用 extends 和 super 关键字）

Type parameters can use "extends" keyword to limit the scope of the actual data types, such as <T extends Number>, where extends does not represent inheritance, but represents the scope of the actual data type, represents T ≤ Number, all the data type under the Number class, such as Integer, Float,..., can be the actual data type, but can't be String, Test and so on.

For instance < T extends Collection>, means that the types of the generic parameters extend to all classes that implement the Collection interface or its sub interfaces. If <String>, the program will compile incorrectly.

类型参数可以使用 extends 关键字来限制实际数据类型的范围，例如 <T extends Number>，其中 extends 不代表继承，而代表 T 的类型不能超过 Number，T ≤ Number 表示 Number 类及其子类的所有数据类型，如 Integer、Float 等，但不能是 String、Test 等。

例如 <T extends Collection>，表示该泛型参数 T 的实际数据类型只能是所有实现了 Collection 接口或者其子接口的类。如果是 <String>，则程序会出现编译错误。

"super" keyword just alike the use of the "extends" keyword for limiting the scope of the actual data types, but in the opposite direction. The "super" is a constraint that generic parameters can only be replaced with its super classes. For instance:

<T super List>

It indicates that the actual data type can only be the List related class and super of List, such as Collection related classes.

使用 super 关键字的作用与 extends 一样，都是限制泛型参数的适用范围，但方向相反。super 是限制泛型参数只能被其父类替换。

7.1.4　Wildcard in type parameters（类型参数里的通配符）

Type parameters can also use wildcard. The purpose of using wildcards is to eliminate the disadvantage that the generic parameters cannot be determined dynamically by the object.

类型参数还可以使用通配符。使用通配符的目的是为了弥补泛型参数不能动态地根据对象来确定类型的不足。

For instance:

```
public class Sample <T extends S> {…}
```

Assume A, B, C,..., Z all the 26 classes implement the S interface. We need to use these 26 generic class parameters when we use them. We have to:

```
Sample<A> a=new Sample<A>();
Sample<B> a=new Sample<B>();
…
Sample<Z> a=new Sample<Z>();
```

This is obviously redundant, so use Object instead of generics. The best way is to use wildcards when creating objects. For instance:

```
Sample<? extends S> sc=new Sample();
```

List<?>represents unbounded wildcard, means <? extends Object>. You can always read Object, but you can't write anything to a list.

List<?> 代表无界通配符，与 List<? extends Object> 相同。可以随时读取 Object，但不能向列表中写入任何内容。

7.2　Generic interfaces（泛型接口）

Generics also work with interfaces.

```
public interface Generator<T> {T next();}
```

The return type of next( ) is T. Using generics for interfaces is the same as for classes.

7.3　Generic methods（泛型方法）

You can also parameterize methods within a class. The class itself may or may not be generic, which is independent of whether you have a generic class or not.

A generic method allows the method to vary independently of the class. You should use generic methods "whenever you can". That is, if it's possible to make a method generic rather than the entire class, it's probably going to be clearer to do so.

To define a generic method, you simply place a generic parameter list before the return type of the method head, like:

```
public <T>T testGen(boolean b, T first, T second) {
   return b ? first:second;
}
…
String s=testGen(true, "a", "b"); //invoke method
Integer i=testGen(false, new Integer(1), new Integer(2));
String k=testGen(true, "pi", "new");
```

可以只将类里的方法定义为泛型方法，泛型方法与方法所属的类是否是泛型类无关。可以只定义泛型方法，而不必定义整个类为泛型类。在方法的返回类型前面放置泛型参数，方法即为泛型方法。

【Example 7-4】**Define a generic method**

```
public class Test {
  public <T>void fun(T x) {
    System.out.println(x.getClass().getName());
  }
  public static void main(String[] args) {
   Test gm=new Test();
     gm.fun(""); gm.fun(1); gm.fun(1.0); gm.fun(1.0F); gm.fun('c'); gm.fun(gm);
  }
}
```

Output：

java.lang.String
java.lang.Integer
java.lang.Double
java.lang.Float
java.lang.Character
Test

7.4 Collection classes（集合类）

When batch data is pracessed, you won't know the quantity or even the exact type of the objects you need. Most languages provide some solutions to solve this essential problem. Java library provides container classes to hold and deal with objects.

Usually, an array is the most efficient approach to hold a group of objects. But an array has a fixed size, and limits its use, because you won't know at the time you're writing the program how many objects you're going to need. The container classes in java.util package needn't to tell how big a container should be.

批量处理数据时通常不知道确切的对象数目，甚至不知道对象的类型，多数程序设计语言都会给出解决办法。Java 类库里提供了容器类，用来保存与处理对象。

通常用数组处理批量数据，但数组的长度固定，限制了其使用，因为在编写代码时，并不知道所需的对象数目。Java.util 包里包含集合类（容器类），不需要指出要处理的数据元素的数目。

7.4.1　Concept of collection (container) classes ［集合（容器）类的概念］

The basic types of container are interfaces including Collection, List, Set, Queue, and Map. These types of interfaces are also known as collection, however, the name Collection is used here as a data type, so the more inclusive term "container" is used.

Containers provide many methods to hold your objects. Unlike arrays, Java container classes will automatically resize themselves. So you can put in any number of objects and don't need to worry about how big to make the container while you're writing a program.

All the top types of collection are generic interfaces.

容器的基本类型是接口，包含 Collection, List, Set, Queue, 和 Map，又称集合（Collection），但 Collection 在这里作为一种数据类型已经被使用，所以把集合叫做容器，容器的含义比集合更为宽泛。

集合提供了很多保存与处理对象的方法。不同于数组，Java 能够自动改变集合的大小。因此在写程序时可以将任意数目的对象放入集合中，而不用担心集合是否能容得下它们。

集合的所有上层类型都是泛型接口。

7.4.2　The hierarchy of the collection framework（集合框架的层次结构）

There are two distinct categories in the Java container library, collection and Map. Containers can only hold objects.

Figure 7-1 is the hierarchy of the collection, the dotted blocks represent interfaces and the solid blocks represent classes. The dotted lines with hollow arrows indicate the inheritance of the interfaces or a particular class implements an interface, and the solid lines with hollow arrows indicate the class inheritances.

Java 容器库中有 2 种容器，Collection 和 Map，容器中只能存放对象。

集合框架的层次结构如图 7-1 所示，其中虚线框表示接口，实线框表示类，带有空心箭头的虚线表示接口的继承关系或者某个类实现了某个接口，带有空心箭头的实线表示类的继承关系。

Chapter 7　Generics and Collections（泛型与集合）

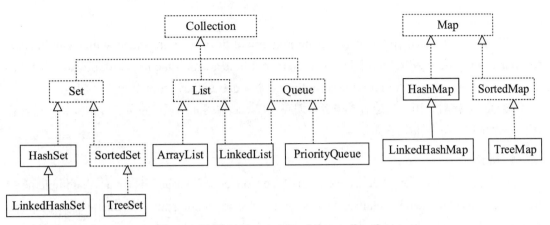

Figure 7-1　Hierarchy of the collection

1. Collection

The classes that implement the List, Set and Queue under the Collection are used to hold a sequence of individual elements. A lot of proctical methods are provided for dealing with the elements.

Queue 的类来存储一系列单位元素，类里有很多实用方法用于处理容器里的对象。

1) Set

Under the Set, there are three classes HashSet, TreeSet and LinkedHashSet that can be used to create containers to hold objects. Elements hold in the Set can't duplicate, they can be accessed to by value. The elements in the Set are sorted or not depend on its two subclasses.

(1) Elements are sorted in TreeSet by value.

(2) Elements are unsorted in HashSet and the storing order of the elements is independent of the entering sequence.

利用 Set 下的三个类 HashSet，TreeSet 和 LinkedHashSet 可以创建容器来存储对象。set 里的元素不能重复，可以通过值访问。Set 里的元素是否排序取决于它的两个子类。

（1）TreeSet 里的元素按值排序。

（2）HashSet 里的元素不排序，并且元素的存放次序与元素进入容器的顺序无关。

2) List

Under the List, there are two classes ArrayList and LinkedList, both of them can be used to create containers to hold objects. The difference between the two is that a LinkedList container object is used to deal with the linked list and provide more operations than the ArrayList container object. The elements hold in the List can duplicate, they are stored in the entering order.

list 下的两个类 ArrayList 和 LinkedList 可以用来创建容器来存储对象。它们的区别是 LinkedList 容器对象用来处理链表序列并且比 ArrayList 容器对象提供更多的操作。list 中元素的值可以重复，元素按它们进入容器的次序排列。

3) Queue

Classes LinkedList and PriorityQueue implement the Queue interface, which can be used to create containers to hold objects. The elements hold in the Queue can duplicate, they are stored in the entering order and output in the same order as they store.

类 Queue 接口，用于创建容器来存储对象。Queue 中的元素可以重复，元素按它们进入容器的次序排列，按存储次序输出。

2. Map

Map is an interface too, but is dependent of the Collection interface. The class that implements Map can be used to create containers to hold objects by key-value pairs. The elements in Map can be accessed to a value by a key, also can be accessed to an object by another object.

Different elements can have same values, but the key must be unique for different elements. Except for the LinkedHashMap, the order of key-value pairs for elements is independent of the order that they enter in the container. The ordering rules are as below.

(1) Elements are unsorted in HashMap

(2) Elements are sorted in TreeMap by key value

(3) Elements in LinkedHashMap are in the entering order

Map 也是接口，但独立于 Collection。实现了 Map 的类可用于创建容器来保存键 – 值（key-value）对对象，Map 中的元素可以按元素的键 key 来访问其值（value），也可以通过一个对象访问另一个对象。

不同元素的值可以相同，但 key 值唯一，不重复。除了 LinkedHashMap，元素中键 – 值对的顺序与它们进入容器的顺序无关。元素在容器里的排序规则如下：

(1) HashMap：不排序。
(2) TreeMap：按 key 值排序。
(3) LinkedHashMap：按进入容器的顺序排序。

7.5　List（列表）

All members in the collection framework are generics. In the collection framework, there are type parameters in Collection, Map, List and Set, and the specific formats are as below:

Collection: interface Collection<E> extends Iterable<E>

Map: interface Map<K,V>

List: Interface List<E> extends Collection<E>

Set: interface Set<E> extends Collection<E>

They all provide many methods for container elements dealing.

1. Methods of List

The List interface adds a number of methods to Collection that allow insertion and removal of elements in the middle of a List.

(1) boolean add(Object o): Add an element to a list, the element type can be any reference type.

For instance:
```
LinkedList<String> mylist=new LinkedList<String>();
mylist.add("aaa");
mylist.add("bbb");
```

(2) boolean addAll(Collection c): Add an array of elements.

(3) Object[] toArray(): Convert and store container elements in an array.

For instance:
```
Object s[]=ay.toArray(); //ay is List object
for(int i=0;i<s.length;i++) System.out.println(s[i]);
```

(4) void clear(): Clear a container.

(5) boolean remove(Object o): Delete an element.

For instance:
```
ay.remove("aaa");
```

(6) boolean removeAll(Collection c): Delete all elements in container c.

(7) boolean retainAll(Collection c): Delete all elements except the elements in container c.

(8) boolean contains(Object o): Judge whether or not object o is in the container?

(9) boolean containAll(Collection c): Judge whether or not all objects in c are in the container?

(10) boolean isEmpty(): Judge whether or not the container is empty?

(11) boolean add(int index, Object o): Add an element at index.

(12) boolean addAll(int index, Collectin c): Add all the elements in c at index.

(13) Object get(int index): Return the element at index.

(14) Object remove(int index): Delete the element at index.

(15) Objetct set(int index, Object o): Replace the element with o at index.

(16) int indexOf(Object o): Return the index of the first element.

(17) int lastOf(Object o): Return the index of the last element.

2. Create a List

The only place where you specify the precise type that you are going to use is at the time of the container creation. So you can create a List like:
```
List<A> ali=new ArrayList<A>();
```

Thus, you get a container object. Notice that the List reference ali, refers to an ArrayList object.

【Example 7-5】 **Create a List without an actual type**
```
import java.util.*;
```

```
public class Test {
  public static void main(String[] agrs) {
    ArrayList al=new ArrayList();
    //Collection<Integer> cal=new ArrayList<Integer>()
    al.add(new Integer(11));
    al.add(1,12);            //add(int,Object)
    al.add(13); //add(Object)
    al.add("Hello");         //or al.add(new String("Hello"));
    System.out.println("Output sequentially");
    for(Object o:al) {       //Object by default
      System.out.println(o);
    }
  }
}
```

Output:

Outputsequncially

11

12

13

Hello

In example 7-5, An ArrayList object is created without actual type, the elements inserted into the collection have no type checking, and the elements retrieve from the container need to be cast. Such as:

```
List a=new ArrayList();
a.add("haha");
a.add(new Cat());
a.add(new Dog());
String s=(String)a.get(0); //get the first element
```

3. Print containers

The default printing behavior (each container's toString() method) produces reasonable results. Elements in Collection are printed surrounding by square brackets, with each element separated by a comma. A Map is surrounded by curly braces, with each key and value associated with an equal sign (keys on the left, values on the right), each pair is separated by a comma.

默认的输出行为（每个容器里的 toString() 方法）都能够恰当地输出容器的元素。Collection 的元素用方括号括起来，彼此用逗号隔开。Map 里的元素用大括号括起来，形式如 key= 值（键在左，值在右），每对用逗号隔开。

【Example 7-6】Create a List with String type, and output it with default toString() method

```
import java.util.*;
public class Test {
  public static void main(String[] args) {
```

```
    List<String> pets=new ArrayList<String>();
    pets.add("Rat");
    pets.add("Cat");
    System.out.println("1: "+pets);
    pets.remove("Rat");   //remove
    pets.add("Dog");
    System.out.println("2: "+pets);
    pets.add(0, "Rat");   //insert at an index
    System.out.println("3: "+pets);
    Collections.sort(pets);
    System.out.println("4: "+pets);
    pets.clear();           //remove all elements
    System.out.println("5: " + pets);
  }
}
```

Output:

1: [Rat, Cat]

2: [Cat, Dog]

3: [Rat, Cat, Dog]

4: [Cat, Dog, Rat]

5: []

【Example 7-7】**Create a String type List, override toString() method**

```
import java.util.*;
public class Test {
  public static void main(String[] args) {
    List<Pet> pets=new ArrayList<Pet>();
    Pet dog=new Pet("Fido", 5, 55.6);
    Pet cat=new Pet("Fluffy", 6, 10.3);
    pets.add(dog);
    pets.add(cat);
    System.out.println(pets);
  }
}
class Pet {
  private String name;
  private int age;
  private double weight; //in pounds
  public String toString() {
    return "Name:"+name+"/Age:"+age+"/Weight:"+weight;
  }
  public Pet(String iname, int iage, double iWeight) {
    name=iname;
    age=iage;
    weight=iWeight;
  }
}
```

Output:

[Name:Fido/Age:5/Weight:55.6, Name:Fluffy/Age:6/Weight:10.3]

【Example 7-8】 **Create a List, output with foreach**

```java
import java.util.*;
public class Test {
  public static void main(String[] agrs) {
                        //ArrayList al=new ArrayList();
    Collection<Integer> al=new ArrayList<Integer>();
    al.add(new Integer(11));
    al.add(1,12);          //12 auto boxing to Integer automatically
    al.add(13);
    al.add(89);
    System.out.println("Output sequencially");
    for(Integer o : al) {  //o is Integer type
      System.out.println(o);
    }
  }
}
```

Output:

Output sequencially
11
12
13
89

【Example 7-9】 **Filling a Collection**

```java
import java.util.*;
public class Collect{
  public static void main(String[] args) {
    Collection<Integer> c=new ArrayList<Integer>();
    for(int i=0;i<10;i++)
      c.add(i); //Autoboxing
    System.out.print(c);
  }
}
```

Output:

[0, 1, 2, 3, 4, 5, 6, 7, 8, 9]

Conclusion:

The add() method puts a new element into a container.

add() means "put it in", because List doesn't care about whether elements are duplicate or not.

All collections can be traversed using foreach syntax.

7.6 Queue（队列）

More methods for queue container are as below.

(1) element(): Get the front element of a queue.

(2) Boolean offer(E e): Insert an element to a queue, which is alike add method.

(3) E peek(): Get the front element of a queue, if the queue is empty, return null.

(4) E poll(): Get and remove the front element of a queue, if the queue is empty, return null.

(5) E remove(): Get and remove the front element of a queue.

【Example 7-10】Create, output and empty a Queue

```java
import java.util.*;
public class QTest {
  public static void main(String str[]){
    Queue<String> que=new LinkedList<String>();
    que.offer("first");
    que.offer("second");
  que.offer("third");
    System.out.println("length of the queue:"+que.size());
    System.out.println(que);
  que.offer("fourth");
  System.out.println("length of the queue:"+que.size());
    System.out.println(que);
    while(que.poll()!=null);
    System.out.println("length of the queue:"+que.size());
    System.out.println(que);
  }
}
```

Output:

length of the queue:3
[first, second, third]
length of the queue:4
[first, second, third, fourth]
length of the queue:0
[]

7.7 Set（集合）

The HashSet class implements the Set interface and is optimized for rapid elements dealing. Set is derived from the Collection too, so there isn't more extra functionality than List. Set owns all methods of List, only few methods not in List.

(1) Iterator iterator()

(2) int size()

The Set has its own special behaviors. A Set's membership is based on the "value" of an object.

The elements in the List can be duplicated and stored in the entering order, but each element in the Set container has a unique value and elements are not sorted.

HashSet 类实现了 Set 接口并进行了优化。

Set 也继承了 Collection，因此 List 的方法 Set 都有，Set 只有少数几个 List 没有的方法。

Set 有其特殊的行为。集合的成员关系是基于对象的"值"。

List 里元素的值可以相重，按进入容器的次序存放。而 Set 里元素的值不重复，元素无序。

【Example 7-11】**Create a HashSet container and output it**

```java
import java.util.*;
public class SetInteger {
  public static void main(String[] args) {
    Random rand=new Random();
    Set<Integer> intset=new HashSet<Integer>();
    for(int i=0;i<1000;i++) {
      Integer a=rand.nextInt(20); intset.add(a);
    }
    System.out.println(intset);
  }
}
```

Output:

[0, 1, 2, 3, 4, 5, 6, 7, 8, 9, 10, 11, 12, 13, 14, 15, 17, 16, 19, 18]

【Example 7-12】**Create an ArrayList container, use add method to add 5 strings and output them**

```java
import java.util.*;
public class Test {
  public static void main(String[] args) {
    List<String> ary=new ArrayList<String>();
    ary.add("one");
    ary.add("two");
    ary.add("three");
    ary.add("four");
    ary.add("two");
    System.out.println(ary);
  }
}
```

Output:

[one, two, three, four, two]

【Example 7-13】**Create a HashSet container, use add method to add 5 strings and output them**

```
import java.util.*;
public class Test {
  public static void main(String[] args) {
    Set<String> ary=new HashSet<String>();
    ary.add("one");
    ary.add("two");
    ary.add("three");
    ary.add("four");
    ary.add("two");
    System.out.println(ary);
  }
}
```

Output:

[two, one, three, four]

7.8 Map（映射）

Map has the ability to map objects to other objects. A Map cannot contain the same keys, each key can map only one value. Conceptually, you can think of a List as a Map with numeric keys, but the two are not direct relationed except that they are both defined in java.util.

Map 能实现一个对象到其他对象的映射。Map 中不能包含相同的 key，每个 key 只能映射一个 value。从概念上而言，可以将 List 看作是有数值键的 Map，而实际上除了 List 和 Map 都是在 java.util 中定义外，二者并没有直接联系。

Map provides useful methods:

(1) Object put(Object key, Object value): Add an element to a container, similar to Set and List.

(2) void putAll(Map c): Add all the elements in c to a container.

(3) Set entrySet(): Return all the elements.

(4) Set keySet(): Return all the keys of the key-value pairs.

(5) Collection values(): Return all the values of the key-value pairs.

(6) void clear(): Clear a Map.

(7) Object remove(Object key): Delete the key-value pair according to the key.

(8) Object get(Object key): Get the key-value pair according to the key. It returns null if the key is not in the container.

(9) boolean containsKey(Object key): Judge whether or not the key value of an element equals "key" is in.

(10) boolean containValue(Object value): Judge whether or not the value of an element equals "value" is in.

(11) boolean isEmpty()

(12) int size()

【Example 7-14】**Create a HashMap container**

```
import java.util.*;
public class Test {
  public static void main(String[] args) {
    Map<Integer,String> ary=new HashMap<Integer, String>();
    ary.put(1, "one");
    ary.put(2, "two");
    ary.put(2, "three");
    ary.put(3, "four");
    ary.put(4, "two");
    System.out.println(ary);
  }
}
```

Output:

{1=one, 2=three, 3=four, 4=two}

【Example 7-15】**Count numbers of every key (0 to 19)**

```
import java.util.*;
public class Statis {
  public static void main(String[] args) {
    Integer freq;
    Random rand=new Random();
    Map<Integer, Integer> m=new HashMap<Integer, Integer>();
    for(int i=0;i<1000;i++) {
      int r=rand.nextInt(20);      //Produce a number between 0 and 19
      freq=m.get(r);               //get value of key r
      m.put(r,freq==null?1:freq+1); //correct value of key r
    }
    System.out.println(m);
  }
}
```

Output:

{0=52, 1=41, 2=61, 3=43, 4=45, 5=55, 6=60, 7=57, 8=43, 9=47, 10=51, 11=51, 12=52, 13=47, 14=43, 15=47, 17=54, 16=55, 19=36, 18=60}

The get(key) method returns null if the key is not in the container.

【Example 7-16】**Create a HashMap container again**

```
import java.util.*;
public class PetMap {
```

```
    public static void main(String[] args) {
      Map<String, Pet> petMap=new HashMap<String, Pet>();
      petMap.put("MyCat", new Pet("Molly",2,12.5));//put to container
      petMap.put("MyDog", new Pet("Ginger",5,22.3));
      System.out.println(petMap);
    }
}
class Pet {
  private String name;
  private int age;
  private double weight;                             //in pounds
  public String toString() {
    return "Name:"+name+"/Age:"+age+"/Weight:"+weight;
  }
  public Pet(String iname, int iage, double iWeight) {
    name=iname;
    age=iage;
    weight=iWeight;
  }
}
```

 Output:

{MyCat=Name:Molly/Age:2/Weight:12.5, MyDog=Name:Ginger/Age:5/Weight:22.3}

Exercises

1. Use ArrayList container to hold 1,2,3,5,6,6,7,8,9 and use toString() to output these numbers.

(1) Use "add" method.

(2) Use "addAll" method, create an Integer array, and then call Arrays.asList method.

2．Use HashSet container to hold 2,2,3,3,4,4,5,5,6,6,7,7 and use foreach syntax to output.

(1) Call "add" method.

(2) Call "addAll" method, create an integer array, and then call Arrays.asList method.

3. Create a class named G with an int g that is initialized in the constructor. Define a method named hop() that displays g. Create an ArrayList object and add G's elements to a List. Use for-each syntax to output the List while calling hop().

4. Write a program. The definition of class A is as below.

```
class A {
  int x, y;
  public A(int a, int b){x=a; y=b;}
  public String toString(){ return "Point:"+x+","+y;}
}
```

(1) Create a Map object with key-value pair Integer-A, add 3 pairs, such as map.put(1, new A(6,7)).

(2) Print keys by calling keySet method.

(3) Print values by calling values method.

(4) Print key-value pairs by calling entrySet.

Chapter 8
Exception Handing
（异常处理）

The ideal time to catch an error is at the source program's compile time, before you even try to run it. However, not all errors can be detected at the compile time. Some problems can only be discovered and must be handled at the program running time. It is need to find some formality that allows the originators of the errors to pass appropriate information to a recipient who will know how to handle the errors properly. We call the exceptions happen at the program running time "runtime exceptions". When a program runs into a runtime error, how can we handle the runtime error so that the program can continue to run or terminate gracefully? This is the subject we will introduce in this chapter.

发现错误的理想时间是在编译源程序时，即运行程序之前。但并非所有的错误都能够在编译时检测到。有些错误只能在运行程序时发现，要将错误信息、包含错误原因告知程序的使用者，以便他们做出正确的处理。把在程序运行时出现的异常称为"运行时异常"。程序正在执行时出现了异常，该如何处理？让程序继续执行或者以合适的方式终止是这一章我们要学习的内容。

8.1 Concepts of exception（异常的概念）

8.1.1 What is an exception?（什么是异常？）

The word "exception" means "Some things happen out of expectation." At the point where the problem occurs, you might not know what to do with it, but you must stop, and must figure out what to do. Some time, you don't have enough information in the current context to fix the problem. So you hand the problem out to a higher context where someone is qualified to make proper decision.

An exception is an event, which occurs during the execution of a program, which disrupts the normal flow of the program's instructions.

Error discovery is a fundamental concern for every program you write. The goals for exception handling in Java are to simplify the creation of large, reliable programs using less codes and try to make your application doesn't have an unhandled error. Reporting errors is the only official approach that Java exception handling.

"异常"的意思是"遇到了意想不到的问题"。在问题发生时,你可能不知道怎么办,可你必须处理。有时,你没有足够的信息,处理不了问题,可以把问题交给能够处理、能够做出正确决定的更高级的背景。

异常是事件,发生在程序运行期间,会打断程序的正常执行。

能够发现运行时的错误是编写程序代码的基本要求。Java异常处理机制能够简化大规模、可靠性高的程序的设计。应用程序必须处理所有的错误,Java异常处理的基本方式是报告错误。

8.1.2　How to deal with exceptions?(如何处理异常?)

When an exception does happen, how the exception is handled? The answer is "Create an exception object to report the exception." The exception object contains information about the error, including its type and the state of the running program when the error occurs. Creating an exception object and reporting the information about the exception is called throwing an exception.

Of cause all errors occur within methods of classes, because outside the methods only fields' definition statements exist. Within a method, by special exception dealing scheme, the errors can be caught and the exception objects can be created. At special cases, you even refuse to handle the errors. Example 8-1 is a simple example of division zero. There is no error is reported at the code compilation time, but it doesn't run normally. It throws an ArithmeticException.when the dividing zero statement c=a/b is executed, The program will stop running, and create an ArithmeticException object to report:

"Exception in thread "main" java.lang.ArithmeticException: / by zero at ExceptionDemo1.main(ExceptionDemo1.java:4)".

【Example 8-1】Throws an ArithmeticException

```java
class ExceptionDemo1 {
  public static void main(String args[]) {
    int a=3,b=0,c;
    c=a/b;
    System.out.println("c="+c);
  }
}
```

如果真的发生了异常，如何处理异常呢？答案是通过创建异常对象来报告异常。异常对象要报告相关的异常信息，诸如异常的类型和发生异常时程序的运行状态。把创建异常对象并报告异常信息称之为抛出异常。

当然所有的错误都发生在类的方法里，因为在方法之外只有成员变量的定义。在方法内部，通过特定的异常处理机制，能够捕捉异常并创建异常对象。特殊情况下，甚至不对异常进行处理。例 8-1 是个除零的例子，编译这段代码时并没有错误报告，但程序确实不能正常执行，当执行到除零语句 c=a/b; 时，程序终止执行，创建异常类 ArithmeticException 的对象并报告如下异常信息：

"主线程" main 发生算术异常，在 Java 代码第 4 行出现除零错误"。

Example 8-1 is a bad example to leave the exception dealing task to the others. It's worth checking if dividing zero about to happen. Give you another throwing exception example. Consider a reference tobj, you might want to check if it has been referenced to an object, before trying to call members of the object through it. You can send information about the error to a larger context by creating an object containing the information and "throw" it out of your current context.

```
if (tobj==null) throw new NullPointerException("tobj=null!");
```

You've thrown an exception (create a NullPointerException object), which allows you to abdicate responsibility for thinking about the issue further. However the thrown exception must be dealt by the accepting codes in a method.

In example 8-2, the last statement attempts to access to an array element out of the array bound, and causes the ArrayIndexOutOfBoundsException throwing. running is termirated and error information is reported.

Exception in thread "main" java.lang.ArrayIndexOutOfBoundsException: 4
at ExceptionDemo2.main(ExceptionDemo2.java:4)

One of the most important aspects of exception dealing is that if something bad happens, don't allow a program to continue run along its original path. You can stop running of the program when an exception is detected and tell what went wrong, or force the program to deal with the problem and let running process return to a stable state.

例 8-1 是把异常处理的任务推给别人的例子。程序应该对除数进行检查，看是否会发生除零的问题。下面再给出另外一个抛出异常的例子，有个引用 tobj，在通过它调用对象的成员之前，对它进行检查，看它是否已经指向了某个对象，用户可以：

```
if (tobj==null) throw new NullPointerException("tobj=null!");
```

这样，你就抛出了一个异常（创建了一个 NullPointerException 对象）。把异常抛出去之后，就不用再考虑如何处理这个异常了。然而，接收异常的那个方法里必须包含处理这个异常的代码。

例 8-2 的最后一条语句试图访问数组里并不存在的越界数据，引起了数组越界异常，程序的执行被终止，并被告知异常的性质以及中断发生的具体位置在 Java 代码的第 4 行。

异常处理的一个很重要的方面是，如果发生了异常，不允许程序沿着当前执行顺序继续执

139

行。检测到异常就终止程序的执行，并告诉用户发生了什么，也可以对异常进行处理，使程序回到一个稳定的状态。

【Example 8-2】 **Throws an ArrayIndexOutOfBoundsException**

```
class ExceptionDemo2 {
  public static void main(String args []){
    int a[]={1,2,3,4};
    System.out.println("a[4]="+a[4]);
  }
}
```

8.2　Exception classes（异常类）

In Java, all the exception classes are the subclasses of the Throwable class, so the Throwable class is at the top of the exception class' hierarchy. The Throwable class has two subclasses, one is the Error class and the other is the Exception class. Figure 8-1 shows the hierarchical structure of the exception classes. The common methods of the Throwable class are listed in Table 8-1. The definition of the Throwable class is as below.

```
public class Throwable extends Object implements Serializable {
  public Throwable() {
    //Constructs a new throwable object with null as its message
    …
  }
  public Throwable(String message) {
    //Constructs a new throwable object with the specified message
    …
  }
  …
}
```

在 Java 中，所有的异常类都是 Throwable 类的子类，所以 Throwable 类在异常类层次结构的顶部。Throwable 类有两个子类，一个是 Error 类，另一个是 Exception 类。图 8-1 是异常类的层次结构。Throwable 类的常用方法如表 8-1 所示。

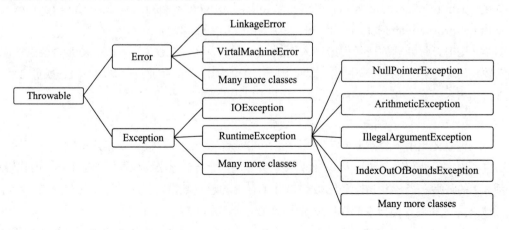

Figure 8-1　Hierarchical structure of the exception classes

Table 8-1　Common methods of Throwable class

Method	Description
getMessage()	Returns the detail message string
getCause()	Returns the cause or null if the cause is non-existent or unknown
toString()	Returns a short description
printStackTrace()	Prints its backtrace to the standard error stream
printStackTrace(PrintStream s)	Prints its backtrace to the specified print stream
printStackTrace(PrintWriter s)	Prints its backtrace to the specified print writer

8.2.1　Error class（Error 类）

System errors are listed right the Error class in Figure 8-1, if one does happen, it will be thrown by the system and we can do nothing. These kinds of errors, such as virtual machine error, main memory overflow, can't be dealt or thrown by our Java codes, which are uncatchable and unrecoverable. If any this kind of errors happens, the running program will be terminated right away.

图 8-1 中的 Error 类的子类对应的是系统内部错误，由系统抛出，Java 代码无法处理。程序代码不能处理或者抛出 Error 类的对象，Error 类的对象如虚拟机出错、内存溢出等，不可捕获、不可恢复。出错时，由系统通知用户并终止程序。

8.2.2　Exception class（Exception 类）

All the subclasses of the Exception class describe errors caused by the program and external circumstances. These errors can be caught and handled, or can be thrown by Java codes at the program running time.

The subclasses of the Exception class can be classified into the unchecked exceptions and the checked exceptions.

Exception 类的子类描述那些由程序本身或者外部环境引起的错误，这些错误能够被捕获，或者由程序代码抛出。

Exception 类的子类可分为非检查异常类和检查异常类。

1. Unchecked exception classes

The RuntimeException class and its subclasses are shown in Figure 8-1, which are used to deal with the exceptions as those classes' names imply, are called unchecked exceptions. The unchecked exceptions are caused by programming logical errors and can't be recovered.

RuntimeException and its subclass exceptions can be thrown automatically. The might be reported information about the exceptions are awful to the program users who are not the program designers. What is more, the only dealing result is stop running the program, sometimes this is not an acceptable decision. To certain situations, the program needs to run, even something happened, instead of stop running. Common unchecked exception classes are listed in Table 8-2.

Of course you can throw an exception by codes like:

```
if (tobj==null) throw new NullPointerException("tobj=null!");
```

The method that includes the above statement, should declare NullPointerException by the throws keyword at the method head. However, the declarations can be omitted for the unchecked exception classes. The compiler doesn't care about whether or not an unchecked exception is dealt with or is declared in a method. That is the name "unchecked" comes from.

The RuntimeException class and its subclasses, the Error class and its subclasses are unchecked exception classes, because the compiler won't force you to deal with these types of exceptions. All the other exception classes that are part of Throwable hierarchy are checked exceptions.

图 8-1 里的 RuntimeException 类及其子类用于处理类名代表的那些异常，这类异常称为非检查异常。非检查异常往往由程序的逻辑错误引起，是不可恢复的。

RuntimeException 类及其子类类型的异常都能够自动抛出。报告的异常信息会让程序的使用者，而非程序的设计者不知所措。特别是，只有一种处理方式，停止程序的运行，有时这是不可接受的，即使程序遇到问题，也不能停止，还要继续运行。常见的非检查异常类如表 8-2 所示。

Table 8-2　Common classes of unchecked exceptions

Exception type	Description
ArithmeticException	Thrown when an exceptional arithmetic condition has occurred
ArrayStoreException	Thrown to indicate that an attempt has been made to store the wrong type of object into an array of objects
ClassCastException	Thrown to indicate that the code has attempted to cast an object to a subclass of which it is not an instance
IllegalArgumentException	Thrown to indicate that a method has been passed an illegal or inappropriate argument
IndexOutOfBoundsException	Thrown to indicate that an index of some sort (such as to an array, to a string, or to a vector) is out of range
NullPointerException	Thrown when an application attempts to use null in a case where an object is required
NumberFormatException	Thrown to indicate that the application has attempted to convert a string to one of the numeric types, but the string does not have the appropriate format

当然你可以像下面这样通过代码抛出异常：

```
if (tobj==null) throw new NullPointerException("tobj=null!");
```

包含这条语句的方法应该在方法的头部，通过 throws 关键字声明 NullPointerException。然而，非检查异常类的声明可以被省略。编译器并不理会代码是否对非检查类型的异常做了处理或者是否声明了该异常。这就是非检查异常类的名字的由来。

RuntimeException 类及其子类和 Error 类及其子类都是非检查异常类，因为编译器不会强迫程序对这类异常进行处理与声明。Throwable 类层次结构中的其余部分都是检查异常类。

2. Checked exception classes

Although a program should handle every abnormal condition, but if you leave an unchecked exception unhandled, the program compiling will not be affected. However, for the checked exceptions, if they are not handled, the program can't be compiled successfully. The checked exceptions are such as IOException, SQLException, ClassNotFoundException and so on. Java compiler forces the programmer to deal with the checked exceptions. Programmers have to write codes to handle the exceptions whether the programmers like or not. Example 8-3 is a checked exception example, which includes a method calling "System.in.read(); ", which may cause an IOException, it is dealt, so the program is compiled successfully. The common checked exception classes are listed in Table 8-3.

程序应该处理所有的异常，但是如果未处理非检查类的异常，程序照样能够通过编译。而对于检查类的异常，如果代码不对其进行处理，程序就不能通过编译。类 IOException、SQLException、ClassNotFoundException 等都属于检查异常类。编译器强迫程序员处理这些异常。不管是否愿意，都必须编写处理这类异常的代码。例 8-3 是检查异常类的例子，例中包含了方法调用语句 System.in.read();，read() 方法可能引起 IOException 类异常，程序做了处理，代码顺利通过编译。常见的检查异常类如表 8-3 所示。

Table 8-3 Common classes of checked exceptions

Exception type	Description
IOException	General class of exceptions produced by failed or interrupted I/O operations
SQLException	An exception that provides information on a database access error or other errors
ClassNotFoundException	Thrown when an application tries to load in a class through its string name using, but no definition for the class with the specified name could be found
FileNotFoundException	An attempt to open the file denoted by a specified pathname has failed
Subclasses of Exception	Defined by programmers

【Example 8-3】 **Catch an exception**

```
import java.io.IOException;
class ExceptionDemo3 {
  public static void main(String args[]) {
    System.out.println("Press Y to end the program.");
    int result='k';
    while(result!='Y'&&result!='y') {
      result=read();
    }
  }
  public static int read() {
```

```
    int result=-1;
    try {
      result=System.in.read();
    }
    catch(IOException e) {
      System.out.println(e.getMessage());
      System.out.println("Fatal error. Ending Program.");
      System.exit(0); //end program
    }
    return result;
  }
}
```

8.3 Catch and deal with an exception(捕获与处理异常)

At run time, if an exception happens, the exception will be thrown automatically and the related information be reported. However this kind of information is terrible for the program users. A good program deals with all possible exceptions.

The checked exception dealing mechanism is a wonderful feature of the Java programming language. you are forced to deal with the checked abnormal conditions, which greatly enhances the program's running reliability. If in a method, a checked exception is thrown, the method must declare what type exception it throws at the method head part. However, if an unchecked exception is thrown in a method, the exception declaration can be omitted. The code that invokes the method must handle the exception or throw it further.

在程序运行时，如果发生了异常，异常会自动抛出并报告相关的异常信息。但这种信息常常让程序的使用者非常困惑。好的程序会处理所有可能的异常。

异常检查处理机制是 Java 程序设计语言的一个非常好的特征，它强迫程序员去处理可能的异常，极大地提高了程序运行的可靠性。如果在一个方法里，包含抛出检查异常的语句，这个方法必须在方法的头部声明被抛出的异常的类型。如果方法里抛出的是非检查类异常，可以省略对该异常类的声明。调用抛出了异常的方法的代码必须处理被抛出的异常，或者进一步抛出这个异常。

How to deal with an exception?

The try…catch…finally structure can be used to catch and handle any kinds of exceptions, checked or unchecked . The try…catch…finally mechanism solves several abnormal conditions in one place means your codes are much easier to write and read because the goal of the codes are not confused with the codes for other purposes.

如何处理异常？

try…catch…finally 结构用于捕捉和处理异常，无论是"非检查"还是"检查"类的异常都可以采用。try…catch…finally 结构专门处理异常，使得异常处理代码与其他功能的代码截然分离，易于代码的编写与阅读。

1. Put suspected things into a try block

Put everything might cause abnormal conditions in a try block, capture and handle all the possible exceptions in one or more catch blocks. The most important step is to enclose the codes that might throw exceptions within a try block. In the "try" block, various suspected method calls and potential problem codes are there. A try block can't be there alone, one or more catch blocks or a final block must fallow it to capture and handle the exceptions.

将所有可能引起异常的语句放到 try 语句块中，然后利用一个或者多个 catch 语句块捕捉与处理异常。把可能抛出异常的代码放到 try 块里是最重要的第一步，try 块里包含各种可能引起异常的方法调用和存在潜在问题的代码。try 语句块不能单独使用，它后面必须跟随一个或多个 catch 语句块，或者一个 final 语句块。

2. Deal with all types of exceptions in the catch blocks

If an exception occurs within the try block, the exception is handled by an exception handler associated with it. To associate an exception handler with a try block, you need to put a catch block next to the try block. Each catch block can detect and handle only one particular type of exception.

If the try block may produce several abnormal conditions, you need to put more catch blocks after the try block. In general, a try…catch structure looks like:

```
try {
  … //may cause N abnormal conditions
}
catch (Exceptiontype1 excp1) {
  … //codes to handle excp1 type of exception
}
catch (Exceptiontype2 excp2) {
  … //codes to handle excp2 type of exception
}
…
catch (ExceptiontypeN excpN) {
  … //codes to handle excpN type of exception
}
```

At the heads of the catch blocks, the exception types (class types) are announced, through the arguments (objects), objects excp1, …, excepN, the exception dealing methods can be called by them.

如果在 try 块中发生了异常，则由与其相关联的异常处理代码处理该异常。要将异常处理代码与 try 块关联，须在 try 块后面放置 catch 块。每个 catch 块只能检测并处理一种特定类型的异常。

在 catch 块的头部，有异常类型（类类型）的声明，异常对象 excp1，…，excpN 可以用来调用处理异常的方法。

Each catch block is an exception handler that handles a particular type of exception indicated by its argument type in the head of the catch block. The exception type that a handler can handle must be the standard types under the Exception class, or the user defined subclasses of the Exception class.

A catch block contains codes that are executed and when an abnormal condition appears in the try block and the abnormal type is in accordance with the catch's argument type. The thrown object from the try block can legally be assigned to an exception handler's argument.

If an abnormal situation happens, the thrown exception from the try black is compared with the type of each argument of the catch blocks one by one, until a matching type is found, or compared all, therefore, the order of the catch blocks should be from special to general. Never arrange catch blocks like example 8-4, the codes in the second catch block and after the second catch blocks would never be executed, because they are the subclasses of Exception class. Example 8-5 is an example of exception types from special to general.

If a father exception type is the catch's argument type, the catch block can deals several exceptions of its subclasses, that is, the catch block can cope with multi types of exceptions. The exception type of a catch block should be selected properly.

每个catch块都包含异常处理代码，处理在catch头部的参数类型指定的特定类型的异常。异常类型必须是Exception类之下的标准异常类，或者是用户自定义的Exception类的子类。

当try块里出现异常，并且异常的类型与catch块的异常类型匹配，则对应的catch块的异常处理代码被执行。抛出的异常对象被赋值给catch的参数。

如果try块抛出异常，该异常依次与每个catch块的参数类型比较，直到找到类型相匹配的catch语句或者比较完了所有catch的参数类型，因此catch块的异常类型的排列顺序应该是从特殊到一般。千万不要像例8-4那样排列catch块，第二个catch及第二个以后的catch里的代码永远不会被执行，因为其异常类型是Exception类的子类。例8-5是异常类型从特殊到一般的例子。

如果catch块的参数类型是一些异常类的共同父类，那么可以用一个catch块处理多个其子类的异常，要根据具体情况来选择catch语句的异常处理类型。

【Example 8-4】 **Bad example, exception from general to special**

```
…
try {
    …
}
catch(Exception e) {
    …
}
catch(ArrayIndexOutOfBoundsException et1) {
    …
}
catch(NumberFormatException et2) {
    …
}
```

【Example 8-5】 **Exceptions from special to general**
```
import java.io.*;
public class ExceptionDemo5 {
  public static void main(String[] args) {
    int i;
    FileReader fr;
    try{
      fr=new FileReader("f:\\abc.txt");
      while((i=fr.read())!=-1)
        System.out.print((char)i);
      fr.close();
    }
    catch(FileNotFoundException e1) {
      e1.printStackTrace();
    }
    catch(IOException e2) { //IOException is the father class of FileNotFoundException
      e2.printStackTrace();
    }
  }
}
```

3. Finally block always is executed

The codes in a finally block are guaranteed to execute, whether any exception happens or not. The finally block is useful for more than just exception handling. It allows a programmer to put clean up codes in a finally block, for instance codes for closing opened files, restoring resources. Example 8-6 is a finally block example, in which the codes in the finally block are always executed.

无论是否发生异常，finally 块里的代码一定被执行。finally 块的作用不仅仅限于作为异常处理机制的一个部分。它让程序员在其中放入一些清理代码，比如关闭文件、恢复资源等。例 8-6 是一个包含了 finally 块的例子，例中 finally 块里的代码总是被执行。

In general, a try…catch…finally block looks like:

```
try{
  … //codes may throw exceptions
}
catch(exception type) {
  … //codes dealing an exception
}
finally {
  … //codes always are executed
}
```

【Example 8-6】 **try…catch…finally**
```
import java.io.*;
public class ExceptionDemo6 {
  public static void main(String[] args) {
```

```
    String s="hello";
    int num;
    try {
      num=Integer.parseInt(s);
      System.out.println(num);
    }
    catch(NumberFormatException e) {
      System.out.println("Input a number!");
    }
    finally {
      System.out.println("This is executed always");
      System.out.println(s);
    }
  }
}
```

 Output:

Input a number!
This is executed always
hello

Change String s="hello"; to String s="123";The results turn to:

 Output:

123
This is executed always
123

8.4　Throw an exception（抛出一个异常）

A method can declare exceptions with the throws keyword and leave the might be thrown exceptions without dealing in the method. If abnormal situations do happen in a method that declares the exceptions, the method can do nothing, just leave the exception dealing task to the method caller. Running example 8-7, an ArithmeticException exception does happen in method div, but the method does nothing, the exception is dealt with in the method main. Without the throws clause at method div, the program would not end normally. Example 8-8 is an example of multiple exception declarations, although the exceptions do happen, but the program ends properly.

You can also throw an exception too with another keyword "throw". You probably noticed that the Java class library provides numerous exception classes. All the classes are descendants of the Throwable class, these various types of exceptions might occur during the program execution and can be thrown in a method with the throw keyword.

If a method encounters an abnormal situation, which the method can't or doesn't like to cope with, the method can throw the exception and let the method caller deal with the thrown exception. For example, a car may break down when it is running, and the car itself can't handle it. Let the driver handle it.

对于方法里可能发生的异常，可以不在方法里处理，但要通过关键字 throws 声明这个异常。如果在声明了异常的方法里真的发生了异常，方法可以什么都不做，让该方法的调用代码去处理异常。例 8-7 的代码运行时，div 方法里发生了算术异常，方法并未对其处理，而是由调用该方法 main 方法对其进行处理，如果 div 方法未对异常类型进行声明，该程序不会正常结束。例 8-8 是通过 throws 声明多个异常类的例子，尽管程序运行中出现了异常，但程序能够正常结束。

也可以使用 throw 关键字抛出异常。你可能注意到了，Java 类库提供了许多异常类，所有的异常类都是 Throwable 类的子类。这些异常类包含了在执行程序期间可能发生的各种类型的异常，它们都可以通过 throw 关键字抛出。

当一个方法里面出现了异常，方法自己没有能力，或者不想处理这个异常，该方法可以将这个异常抛出，交给调用这个的方法的代码去处理。例如汽车在行驶时可能会出现故障，汽车本身没有办法处理这个故障，那就让司机来处理。

1. Declare exceptions with the keyword "throws"

Multiple exceptions can be declared in a throws clause of a method. The syntax of the exception declaration with the throws keyword as below:

```
Methodname() throws Exception1, Exception2, …, ExceptionN {
    …
}
```

【Example 8-7】**Declare the ArithmeticException**

```java
import java.io.*;
class ExceptionDemo7 {
  public void div(int a,int b) throws ArithmeticException {
    System.out.println(a+"/"+b+"="+a/b);
  }
  public static void main(String args []) {
    ExceptionDemo7 e7=new ExceptionDemo7();
    try {
      e7.div(8,2);
      e7.div(5,0);
      e7.div(6,2);
    }
    catch(ArithmeticException e) {
      System.out.println("An exception happens here."+e.getMessage());
    }
  }
}
```

Output:

8/2=4

An exception happens here./ by zero

【Example 8-8】 **Declare the FileNotFoundException and the IOException**

```java
import java.io.*;
public class ExceptionDemo8 {
  public static void main(String args []) {
    ExceptionDemo8 e8=new ExceptionDemo8();
    try {
      e8.copyFile();
    }
    catch(FileNotFoundException e) {
      System.out.println("File can't be found:"+e.getMessage());
    }
    catch (IOException e) {
      System.out.println("I/O Exception imformation:"+e.getMessage());
    }
  }
  public void copyFile() throws FileNotFoundException,IOException {
    FileReader fr=new FileReader("c:\\abc.txt"); //the file does not exit
    FileWriter fw=new FileWriter("d:\\abc.txt");
    int i;
    while((i=fr.read())!=-1) {
      fw.write(i);
    }
    fr.close();
    fw.close();
  }
}
```

Output:

File can't be found:c:\abc.txt

2. Throw an exception with the keyword "throw"

If abnormal situations happen, you may also throw the potential exceptions with the "throw" keyword and keep the might be exceptions unhandled in a method, but the caller must handle it. A method that includes the throw statements, should declare the types of the thrown exceptions with the throws clause, but for the unchecked exceptions, the throws clause can be omitted.

A method can throw multiple exceptions, they all should be declared with the "throws" keyword. If a method calls the exception throwing method, and don't want to catch and handle the thrown exceptions, the calling method can throw and declare the exceptions further, until method main. Even the main method can refuse to deal with the exceptions and declare them with throws clause. That's a terrible idea, make sure deal with the exceptions at the right time. The thrown exceptions should be captured and handled use try… catch schemes in the calling method. Example 8-9 is an example of throwing an exception in a method and the caller method main deals with it.

　　如果发生异常，用户也可以不给予处理，利用 throw 关键字抛出异常，将异常处理任务留给方法的调用代码。包含抛出异常语句的方法要用 throws 声明所抛出异常的类型，但对于"非检查"异常类，throws 异常声明可以省略。

　　一个方法可以抛出多个异常并用 throws 关键字声明所有被抛出的异常。如果一个方法调用了抛出异常的方法，但并不想对异常进行捕捉与处理，可以进一步抛出，一直到 main 方法。main 方法也可以继续用 throws 声明异常，而不处理异常。这不是正确的做法，要在恰当的时候处理异常。

　　调用抛出了异常的方法时必须使用 try…catch 语句捕获与处理异常。例 8-9 是在一个方法里抛出异常，由调用这个方法的 main 处理异常的例子。

【Example 8-9】**Throw and catch an exception**

```
import java.io.*;
import java.util.Scanner;
class ExceptionDemo9 {
  public static void main(String args []) {
    ExceptionDemo9 e9=new ExceptionDemo9();
    try {
      e9.checkScore();
    }
    catch (Exception e) {
      System.out.println("Here is an abnormal:"+e.getMessage());
    }
  }
  public void checkScore() throws Exception {
    int score;
    Scanner scanner=new Scanner(System.in);
    score=scanner.nextInt();
    if(score<0||score>100)
      throw new Exception("You entered an wrong score!");
    else
      System.out.println("You score:"+score);
  }
}
```

Input:

105

Output:

Here is an abnormal:You entered an wrong score!

8.5 Define your own exceptions（用户自定义异常类）

Although Java library provides us many exception classes, but can't foresee all the errors you might meet with, so you can define your own exception classes for your special problems that your application might encounter. The custom defined exception classes must extend the Exception class, that is, the custom defined exception classes must be the subclasses of Exception class.

尽管 Java 类库中提供了很多异常类，但不可能预见到所有的异常，用户可以针对特定的应用程序可能遇到的特定问题，定义自己的异常类。用户定义的异常类必须是 Exception 类的子类。

How do you create your own exceptions?

(1) Define a class that inherits the Exception class or its subclasses.

(2) Add a constructor method.

(3) Use the throw keyword to throw an exception object in a method.

(4) Capture and handle exceptions in another method.

Example 8-10 is a custom defined exception class example.

如何定义自己的异常？

（1）定义一个异常类，该类继承 Exception 类或者 Exception 类的子类。

（2）为该异常类添加构造方法。

（3）在该类的一个方法中使用关键字 throw 抛出异常。

（4）在该类的另一个方法中捕获并处理异常。

例 8-10 是用户自定义异常类的例子。

【Example 8-10】Define your own exceptions

```java
class BankException extends Exception { //Define a BankException class
  String message;
  public BankException(int m,int n) {
    super();
    message="Account balance error:in:"+m+"  out:"+n;
  }
  public String reMess() {
    return message;
  }
}
class Bank {                            //define a Bank class
  private int money=0;
  public void balance(int m,int n)throws BankException {
    if( m<0||n>0||m+n<0)                //m:deposit, n: withdraw
      throw new BankException (m,n);    //throw a BankException exception
    int netBalance=m+n;
```

```
      System.out.println("balance:"+m+" out:"+n+"  netBalance:"+netBalance);
      money+=netBalance;
    }
    public int getMoney() {
      return money;
    }
}
class TestCustomException {
    public static void main(String args []) {
      Bank bank=new Bank();
      try{
        bank.balance(100, -100);
        bank.balance(400, -300);
        bank.balance(200, -300);
        bank.balance(200, -100);
      }
      catch(BankException e) {
        //e.printStackTrace();
        System.out.println(e.reMess());
      }
      System.out.println(bank.getMoney());
    }
}
```

Output:

balance:100 out:-100 netBalance:0
balance:400 out:-300 netBalance:100
Account balance error：in:200 out:-300
100

Exercises

1. Compile and run the program below, and then encloses.rs(-92); with try... catch structure, run it again.

编译、运行下面这段程序，然后用 try...catch 包围 ts.rs(-92); 语句，再运行。

```
import java.io.*;
class TryThrows {
    double radius;
    public static void main(String args[]) {
      TryThrows ts=new TryThrows();
      ts.rs(-92); //enclose with try… catch structure
    }
    public void rs(double newRadius) {
      if (newRadius>=0)
        radius=newRadius;
```

```
        else
            throw new IllegalArgumentException("Radius can't be negative");
    }
}
```

2. *Write a program. Input a character from the keyboard. Check and print the input, until the input is a "Q".*

写程序,输入一个字符,检查该字符,直到输入的字符是"Q"。

(1) Use try... catch structure. 使用 try... catch 结构。

(2) Use throws. 使用关键字 throws。

3. *Write a program. Design a method "String getline()"(use System.in.read()), which receives and returns a string, the main method calls the getline() and prints the string returned from the getline() method.*

写程序,定义一个方法 getline()(利用 System.in.read()),该方法接收并返回一个字符串。在主方法中调用该方法,输出从该方法返回的字符串。

Chapter 9
I/O（输入/输出）

Java Input/Output (I/O) deals with reading data from a source and writing data to a destination. Both of the source and destination can be a hardware device such as the keyboard, or software programs such as two running programs communicate each other.

9.1　Concept of I/O stream（输入/输出流的概念）

A stream is a flow of data from a source to a destination. Typically, your program will be one end of that stream and some other node (for instance, a file) will be the other end. An object associated with the source is called the input stream and an object associated with the destination is called the output stream. You can read data from the input stream, but you can't write data to it. Conversely, you can write to an output stream, but you can't read data from it.

A program reads the data in the source from the input stream. When a program needs to read data, it opens an input stream that associates with the data source, which can be a file、memory or network connection. When a program needs to write (output) data, just write data to the output stream that associates with the destination where the data is going to. The program passes the information to the destination by writing data to the output stream. When a program needs to write data, it opens an output stream to its destination.

Java supports two types of streams: byte and character. Input and output of byte data are handled by InputStream and OutputStream classes. Input and output of character data are handled by Reader and Writer classes. Figure 9-1 shows the input stream and the out stream.

流是数据从数据源到另一端（终端）的数据的流动。比较典型的情形是，数据流的一端是程序，另一端是其他的结点（比如一个文件）。与数据源关联的对象是输入流，与数据终端关联的对象是输出流。可以从输入流中读取数据，但不可以写入数据。相反地，可以向输出流写入数据，但不能读取数据。

程序从输入流中读取数据源的数据。当程序读取数据时，首先打开数据源的输入流，数据源可以是文件、内存或者是网络连接。程序输出数据时要指出数据的目的地，将数据写入输出流。程序写数据时，要打开通向目的地的输出流。

Java 支持两种流：字节流和字符流。支持字节输入 / 输出流的类是 InputStream 类和 OutputStream 类，支持字符输入 / 输出流的类是 Reader 类和 Writer 类。图 9-1 表示了输入流和输出流。

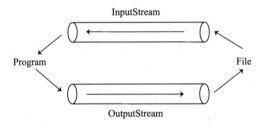

Figure 9-1　InputStream andOutputStream

No matter what type of data or wherever the data to and from, the data input and output almost include the same tasks:

(1) Open a stream

(2) Read/write data

(3) Close stream

无论什么类型的数据，无论数据来自哪里，流向哪里，数据的输入和输出都包含 3 个步骤：

（1）打开流；

（2）读 / 写数据；

（3）关闭流。

9.2　Byte stream（字节流）

Byte streams deal with the data in bytes. The byte streams are used to deal with binary data, such as images and sounds.

Byte input is realized by subclasses of the InputStream class and byte output is realized by subclasses of the OutputStream class. The InputStream class and the OutputStream class are the basic classes of stream classes, all the other byte stream classes are direved from them. Figure 9-2 shows the hierarchical structure of the byte stream classes.

字节流以字节为单位处理数据，字节流用于处理二进制数据，如图像和声音。

字节输入由 InputStream 类的子类实现，字节输出由 OutputStream 类的子类实现。InputStream 类和 OutputStream 类是字节流类的基类，其他的字节流类都是它们的子类。图 9-2 是字节流类的层次结构。

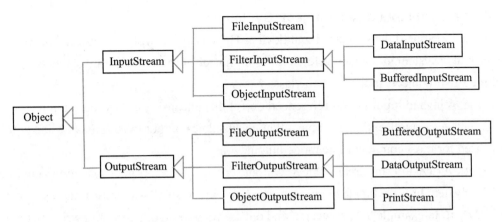

Figure 9-2 Hierarchical structure of the byte stream classes

9.2.1 InputStream/OutputStream（InputStream/OutputStream 类）

1. InputStream

The InputStream class is an abstract class, and it is the super class of all classes dealing with byte input streams. Byte input is realized by subclasses of the InputStream class. The InputStream class provides various methods for dealing with the byte input. Table 9-1 shows the common methods of the InputStream class.

InputStream 类是一个抽象类，它是其他输入字节流相关类的父类。字节输入由 InputStream 类的子类实现，InputStream 提供了处理字节输入的各种方法，InputStream 类的常用方法如表 9-1 所示。

Table 9-1 Common methods of InputStream class

Method	Description
read()	Reads the next byte of data from the input stream
read(byte[] b)	Reads some number of bytes from the input stream and stores them into array b
read(byte[] b, int off, int len)	Reads up len bytes of data from the input stream into an byte array.
skip(long n)	Skips over and discards n bytes of data from the input stream
available()	Returns an estimate of the number of bytes that can be read (or skipped over) from the input stream without blocking by the next invocation of a method for the input stream
close()	Closes the input stream and releases any system resources associated with the stream
reset()	Repositions the stream to the position at the time the mark method was last called on the input stream
markSupported()	Tests if the input stream supports the mark and reset methods

2. Standard input stream object: System.in

We call keyboard the standard input device. There is an input stream object that we are so famiLer with "System.in", which references to an object of the InputStream's subclass. There are three stream objects in the System class.

(1) in: A java.io.InputStream referenced object that encapsulates keyboard input.

(2) out: A java.io.PrintStream object that encapsulates output to a command-line window and is used for most command line mode output.

(3) err: A java.io.PrintStream object. Output sent to System.err goes to a command line.

The System.in is an InputStream referenced to object, so all methods in table 9-1 can be invoked by it. In example 9-1, objects in and out are used to invoke read and write methods.

键盘是标准输入设备。System.in 是我们熟悉的 InputStream 指向的 InputStream 子类的对象。System 类里定义了 3 个流类的对象。

(1) in：java.io.InputStream 指向的对象，它封装了键盘输入。

(2) out：java.io.PrintStream 对象，它封装了命令行输出。

(3) err：java.io.PrintStream 对象，其输出也是到命令行。

System.in 是对象，通过它可以调用表 9-1 的所有方法。例 9-1 通过 in 和 out 对象调用 read 和 write 方法。

【Example 9-1】 **Copy keyboard input character to the screen**

```java
import java.io.*;
public class ReadBin {
  public static void main(String[] args) {
    int ch;
    InputStream in=System.in;
    OutputStream out=System.out;
    try {
      while((ch=in.read())!=-1) {
        out.write(ch);
      }
      in.close();
      out.close();
    }
    catch(IOException e) {
      error("I/O Exception:"+e);
    }
  }
  public static void error(String er) {
    System.err.print("Copy:"+er);
    System.exit(1); //non-zero, tell system something wrong
  }
}
```

Input :

123 45

 Output:

123 45

3. OutputStream

The OutputStream class is an abstract class too, and it is the super class of all classes dealing with byte output stream. An output stream accepts output bytes and sends them to the destination. Byte output is realized by subclasses of the OutputStream class. It provides various methods for dealing byte output. Table 9-2 shows the common methods of OutputStream class.

OutputStream 类也是一个抽象类，它是所有字节输出流相关类的父类。输出流将字节数据送到数据终端。字节的输出操作由 OutputStream 类的子类实现，OutputStream 提供了许多处理字节输出的方法。OutputStream 类的常用方法如表 9-2 所示。

Table 9-2 Common methods of OutputStream class

Method	Description
write(int b)	Writes the specified byte to the output stream
write(byte[] b)	writes bytes in array b to the output stream
write(byte[] b, int off,int len)	Writes len bytes from the specified byte array starting at offset off to the output stream
void flush()	Flushes the output stream and forces any buffered output bytes to be written out
close()	Closes the output stream and releases any system resources associated with the stream

9.2.2 FileInputStream/FileOutputStream（FileInputStream/FileOutputStream 类）

1. FileInputStream

The FileInputStream class is the subclass of the InputStream class. A byte file input is realized by this class. Before dealing with a binary file, a FileInputStream object (input stream) must associate with an external file. All the methods in the FileInputStream class are inherited from its super class. This class is appropriate for simple file reading.

A FileInputStream object (input stream) obtains input bytes from the file in a file system. FileInputStream is meant for reading streams of raw bytes such as image data. Prototypes of the two constrctors of the FileInputstream class are as below.

FileInputStream 是 InputStream 的子类，该类实现字节（二进制）文件的读取。处理二进制文件输入之前，必须先将 FileInputStream 对象（输入流）与外部二进制文件相关联。FileInputStream 中的所有方法都来自它的父类，该类适用比较简单的文件读取。

FileInputstream 对象（输入流）从文件系统的文件获取输入字节。FileInputstream 用于读取原始字节流，如图像数据。FileInputstream 类的 2 个构造方法的原型如下。

```
public class FileInputStream extends InputStream {
  public FileInputStream (String name) throws FileNotFoundException {
    //Creates a FileInputStream object by opening a connection to an actual file "name", the file name contains the file path.
    …
  }
  public FileInputStream (File file) throws FileNotFoundException {
    //Creates a FileInputStream object by opening a connection to an actual file, the "file" is a File object.
    …
  }
  …
}
```

Example 9-2 and 9-3 read file Hello.java that is the same with file CountCap.java, but takes different name and stays at f disk. Example 9-4 reads a group of bytes from file aa.dat, before running example 9-4, example 9-5 and 9-6 need to run first to write data to file aa.dat.

例 9-2 和 9-3 读取 Hello.java 文件，Hello.java 跟 CountCap.java 文件的内容相同，只是文件名不同，位于 f 盘下。例 9-4 读取文件 aa.dat，执行例 9-4 前需要先执行例 9-5 和 9-6 向 aa.dat 里写入数据。

【**Example 9-2**】**Count characters and capital letters of Hello.java, which content is the code below**

```
import java.io.*;
class CountCap {
  public static void main(String[] args) throws IOException {
    int ch, total, capital=0;
    FileInputStream in;
    in=new FileInputStream("f:/Hello.java");
    for(total=0;(ch=in.read())!=-1;total++) {
      if(ch>='A'&&ch<='Z') capital++;
    }
    in.close();
    System.out.println(total+"chars, "+capital+"capitals");
  }
}
```

 Output:

348chars, 16capitals

【Example 9-3】Read from Hello.java

```
import java.io.*;
public class ReadBin {
  public static void main(String[] args) throws IOException {
  FileInputStream s=new FileInputStream ("f:/Hello.java");
    int c;
    while((c=s.read())!=-1)
    System.out.print((char)c);
    s.close();
  }
}
```

【Example 9-4】Read from aa.dat

```
import java.io.*;
public class ReadBin {
  public static void main(String[] args) {
    String fileName="f:/aa.dat";
    int i=0, j, s;
    FileInputStream instr;
    int[] kk=new int[30];
    try {
      File file=new File (fileName);
      instr=new FileInputStream(file);
      while((s=instr.read())!=-1)
        kk[i++]=s;
      instr.close();
    }
    catch(IOException iox) {
      System.out.println("IO Problems with "+fileName);
    }
    for(j=0;j<i;j++) {
      System.out.print(kk[j]);
      System.out.print(",");
    }
  }
}
```

Output:

21,32,0,16,9,67,23,0,13,10,21,32,0,16,9,67,23,0,13,10,

2. FileOutputStream

The FileOutputStream class is the subclass of the OutputStream class. A byte file output is realized by this class. FileOutputStream associates a binary output stream with an external file. All the methods in the FileOutputStream class are inherited from its super class. Before manipulating a file, you must associate the file with a FileOutputStream object. The prototypes of the constructors are as below.

FileOutputStream 类是 OutputStream 的子类，该类用于字节文件的输出。FileOutputStream

将二进制输出流与外部文件相关联。FileOutputStream中的所有方法都来自它的父类。在对文件进行操作之前，须将文件与FileOutputStream对象关联。FileOutputStream类构造方法的原型如下。

```
public class FileOutputStream extends OutputStream {
  public FileOutputStream (String name, boolean append) throws FileNotFoundException {
      //Creates a file output stream, the file name is "name".
      //If the second argument is true, data will append to the end of the file
      …
  }
  public FileOutputStream (File file, boolean append) throws FileNotFoundException {
     //Creates a file output stream, the file represented by the specified File object.
     //If the second argument is true, data will append to the end of the file.
      …
  }
  …
}
```

The constructors initialize the file output stream (object) for a spetified file. If there isn't a file there, a new file would be created. If the file is there, the constructor would empty the file, the data in the file will be lost. To retain and append new data to the file, pass "true" to the second argument while creating an object. In fact, the second argument has a default value "false", if the actual parameter is "true", data will append to the end of the file. if "false", the file will be emptied before the new data.

FileOutputStream类的构造方法初始化指定文件的文件输出流（对象）。如果文件不存在，构造方法会创建一个文件，如果文件已经存在，则清空文件，原有数据将丢失。如果想保留文件里的数据并在原数据后追加新数据，在创建对象时用true做第二个形参的实参。实际上，append参数的默认值是false，表示清空文件后再写入新数据。

【Example 9-5】Empty file if file is not there

```
import java.io.*;
class WriteBin {
  public static void main(String[] args) throws IOException {
    String fileName="f:\\aa.dat";
    byte[] pp={21,32,0,16,9,67,23,0,13,10};
    FileOutputStream dataOut=new FileOutputStream(fileName);
    dataOut.write(pp);
    dataOut.close();
    System.out.println("Writing success!");
  }
}
```

 Output:

Writing success!

【Example 9-6】Append new data if file is there

```
import java.io.*;
```

```
class WriteBin {
  public static void main(String[] args) throws IOException {
    String fileName="f:\\aa.dat";
    byte[] pp={21,32,0,16,9,67,23,0,13,10};
    FileOutputStream dataOut=new FileOutputStream(fileName, true);
    dataOut.write(pp);
    dataOut.close();
    System.out.println("Writing success!");
  }
}
```

 Output:

Writing success!

【Example 9-7】**Copy a file**

```
import java.io.*;
public class CopyFile {
  public static void main(String args []) {
    try {
      FileInputStream input=new FileInputStream("e:\\org.bmp");
      FileOutputStream output=new FileOutputStream("f:\\org.bmp");
      int c;
      while((c=input.read())!=-1) {
        output.write(c);
      }
      input.close();
      output.close();
      System.out.print("Copy success!");
    }
    catch (IOException e) {
      e.printStackTrace();
    }
  }
}
```

 Output:

Copy success!

9.2.3　FilterInputStream/FilterOutputStream（FilterInputStream/FilterOutputStream 类）

　　Filter streams filter bytes for spetial purpose. That is, read bytes and convert them to primitive types of data. The basic byte input stream provides a read method that can only be used for reading bytes. If you want to read integers, doubles, or strings, you need a filter class to wrap the byte input stream. FilterInputStream and FilterOutputStream are the super classes of DataInputStream and DataOutputStream. When need to process primitive types of data, use DataInputStream and DataOutputStream classes.

过滤器类可以根据需要对字节数据进行处理。即先读取字节，再把字节转换成基本数据类型的数据。基本字节输入流类的 read 方法只能读字节，如要读整数、小数、字符串等，需要 FilterInputStream 和 FilterOutputStream 类对字节数据进行包装和处理。FilterInputStream 和 FilterOutputStream 是 DataInputStream 和 DataOutputStream 的父类。DataInputStream 和 DataOutputStream 用于包装和处理基本类型的数据。

9.2.4 DataInputStream/DataOutputStream（DataInputStream/DataOutputStream 类）

DataInputStream allows you to read different types of primitive data as well as String objects. All the methods start with "read", such as readByte(), readFloat() and etc. Along with its companion DataOutputStream, you can move primitive data from one place to another via a stream.

DataInputStream reads bytes from a stream and converts them into appropriate primitive type values or String objects.

DataInputStream extends FilterInputStream and implements the DataInput interface. Table 9-3 shows the common methods of DataInputStream class.

DataInputStream 用于读取基本类型的数据以及 String 对象数据，DataInputStream 的所有方法都以 read 开头，如 readByte()，readFloat() 等。与 DataOutputStream 一起，可以通过数据"流"将基本类型的数据从一个地方移到另一个地方。

DataInputStream 对象从流中读取字节，并将它们转换为基本类型的数据或 String 对象。DataInputStream 类继承 FilterInputStream 类并实现 DataInput 接口。DataInputStream 的常用方法如表 9-3 所示。

Table 9-3　Common methods of DataInputStream class

Methods	Description
read(byte[] b)	Reads number of bytes from the contained input stream and stores them into array b
read(byte[] b,int off,int len)	Reads up len bytes from the contained input stream into array b
readInt()	Reads an int data
readChar()	Reads char data
readByte()	Reads a byte
readBoolean()	Reads boolean

DataOutputStream formats each of the primitive types and String objects onto a stream in such a way that any DataInputStream, on any machine, can read them. All the methods start with "write", such as writeByte(), writeFloat() and etc.

The goal of DataOutputStream is to put data elements on a stream in a way that DataInputStream can portably reconstruct them. DataOutputStream extends FilterOutputStream and implements the DataOutput interface. Table 9-4 shows the common methods of DataOutputStream class.

DataOutputStream 对基本数据类型以及 String 对象进行格式化处理,并将其置入数据"流",以便任何机器上 DataInputStream 都能正常地读取它们。所有方法都以"wirte"开头,如 writeByte(),writeFloat() 等。

DataOutputStream 的作用是将数据置入数据流,以便 DataInputStream 能够方便地重新构造它们。DataOutputStream 继承 FilterOutputStream 类并实现 DataOutput 接口。DataOutputStream 的常用方法如表 9-4 所示。

Table 9-4　Common methods of DataOutputStream class

Methods	Description
write(int b)	Writes the specified byte (the low eight bits of argument b) to the underlying output stream
write(byte[] b,int off,int len)	Writes len bytes from the specified byte array starting at offset off to the underlying output stream
writeBoolean(boolean v)	Writes a boolean to the underlying output stream a 1-byte
writeByte(int v)	Writes out a byte to the underlying output stream a 1-byte
writeChar(int v)	Writes a char to the underlying output stream a 2-byte
writeInt(int v)	Writes an int to the underlying output stream four bytes, high byte first
writeBytes(String s)	Writes out the string to the underlying output stream a sequence of bytes
writeChars(String s)	Writes a string to the underlying output stream a sequence of characters

9.2.5　BufferedInputStream/BufferedOutputStream
（BufferedInputStream/BufferedOutputStream 类）

These two classes provide buffers for input and output operations, support objects buffering their data to avoid every write or read directly to and from a stream. When a buffered stream is created, its buffer size can be specified explicitly or by default, 32 bits.

So far we have beening read and written data one byte at a time. Disk access is much slower than the processing performed in memory. To minimize the number of the disk accessing, Java provides buffering solution.

这两个类为输入和输出数据提供缓冲区,支持对缓冲区数据的读写,避免每次都直接对流读写数据。创建带缓冲区的流时,可以显示地设定缓冲区的大小,也可以采用默认大小,即 32 位。

到目前为止,每次只能读/写一个字节。磁盘的读写速度比内存慢,为了减少访问磁盘的次数,Java 支持缓冲区读写。

1. BufferedInputStream

The BufferedInputStream class adds "mark" and "reset" methods for the ability to buffer the input data. When creating a BufferedInputStream object (a stream), an internal buffer array is created. When bytes from the stream are read or skipped, the internal buffer is refilled as necessary from the contained input stream, many bytes at a time. The mark operation remembers the position in the input stream and the reset operation causes all the bytes since the most recent mark operation to be reread before new bytes are taken from the contained input stream. The prototype of the BufferedInputStream constructor is as below.

BufferedInputStream 类增加了 mark 和 reset 方法，使其能够为输入数据建立缓冲区。创建 BufferedInputStream 对象（流）时，会自动创建一个内部数组，即缓冲区。当从流中读取或跳过字节时，将根据需要从输入流中取出数据填充内部缓冲区，每次可以填充多个字节的数据。mark 操作会记住输入流中读取数据的位置，而 reset 操作会重新读取最近的 mark 操作后读取的数据，然后再从输入流中取数据放入缓冲区。BufferedInputStream 类构造方法的原型如下。

```
public class BufferedInputStream extends FilterInputStream {
    public BufferedInputStream(InputStream in, int size) {
        //Creates a BufferedInputStream with the specified buffer size, and saves its argument, "in", for later use.
        …
    }
    …
}
```

2. BufferedOutputStream

The BufferedOutputStream class implements a buffered output stream. By setting up such an output stream, an application can write bytes to the underlying output stream without necessarily causing a call to the underlying system for each byte written. The prototype of BufferedOutputStream constructor is as below.

BufferedOutputStream 类实现缓冲输出流。通过设置这样的输出流，应用程序向底层的输出设备写入字节，不会每写一个字节就引起底层的系统调用。BufferedOutputStream 类构造方法的原型如下。

```
public class BufferedOutputStream extends FilterOutputStream {
    public BufferedOutputStream(OutputStream out,int size) {
        //Creates a new buffered output stream to write data to the specified underlying output stream with the specified buffer size.
        …
    }
    …
}
```

9.2.6 PrintStream（PrintStream 类）

1. PrintStream

PrintStream class is for formatting the output. The original intent of PrintStream was to print all the primitive type data and String objects in a viewable format, which is different from DataOutputStream.

PrintStream 用于格式化输出。PrintStream 最初的目的是按恰当的格式输出基本类型的数据和 String 对象的数据，这一点与 DataOutputStream 不同。

PrintStream class adds more methods for the ability to print representations of various data conveniently. All characters printed by a PrintStream are converted into bytes, the platform's default character encoding. The PrintWriter class should be used in situations that require writing characters rather than bytes.

The two important methods in PrintStream are print() and println(), which are overloaded to print data of various types. The difference between print() and println() is that the latter adds a newline when it's done. Table 9-5 shows the common methods of PrintStream class. The prototypes of constructors of the PrintStream class are as below.

PrintStream 增加了更多的方法，能够输出各种形式的数据。利用 PrintStream 输出的字符按照平台默认的编码方式转换为字节。如果输出字符而不是字节，则使用 PrintWriter 类。

PrintStream 类的两个重要方法是 print() 和 println()。这两个方法的多个重载方法可以输出所有类型的数据。print() 和 println() 的差异是后者多了一个新行符。PrintStream 类的常用方法如表 9-5 所示。PrintStream 类构造方法的原型如下。

```
public class PrintStream extends FilterOutputStream implements
Appendable, Closeable {
   public PrintStream(File file) throws FileNotFoundException {
     //Creates a new print stream, without automatic line flushing.
     …
   }
   public PrintStream(String fileName) throws FileNotFoundException {
     //Creates a new print stream, without automatic line flushing.
     …
   }
   public PrintStream(OutputStream out) {
     //Creates a new print stream. This stream will not flush automatically.
     …
   }
   …
}
```

Table 9-5　Common methods of PrintStream class

Methods	Description
close()	Closes the stream
print(boolean b)	Prints a boolean value
print(char c)	Prints a character
print(char[] s)	Prints an array of characters
print(int i)	Prints an integer
print(String s)	Prints a string

【Example 9-8】 Randomly print 5 integers (values witnin 100) and write them to file rand.dat

```
import java.io.*;
import java.util.Random;
```

```
public class PFile {
  public static void main(String[] args) {
    PrintStream ps;
    FileOutputStream file;
    int rs;
    try {
      file=new FileOutputStream("f:\\rand.dat");
      Random r=new Random();
      ps=new PrintStream(file);
      for(int i=0;i<5;i++) {
        rs=r.nextInt(100);
        System.out.print(rs+"\t");
        ps.print(rs+"\t");
      }
      ps.close();
    }
    catch(Exception e) {
      e.printStackTrace();
    }
  }
}
```

Output:

67 19   94   41   51

2. Standard output stream object: System.out

The screen is the standard output device. Throughout this book, you've seen how to write to the console (the screen) using System.out, which is already pre-wrapped as a PrintStream object for most command line mode output. System.err is a PrintStream object too. They are the object fields of System class. System.out is an object, all PrintStream methods can be invoked by it. In example 9-9, method println of PrintStream class is invoked by System.out.

显示器是标准输出设备。全书一直在用 System.out 输出，它已被预封装成一个 PrintStream 对象，用于命令行输出。System.err 也是一个 PrintStream 对象，它们都是 System 类的对象成员。System.out 是对象，可以通过它调用 PrintStream 类的方法，例 9-9 中，通过 System.out 对象调用 PrintStream 类的 println 方法。

【Example 9-9】System.out

```
import java.util.Scanner;
public class Out {
  public static void main(String[] args) {
    Scanner input=new Scanner(System.in);
    System.out.println("Enter two integers:");
    int first=input.nextInt();
    int second=input.nextInt();
    System.out.println("You entered "+first+" and "+second);
  }
}
```

Output:

Enter two integers:
12 89
You entered 12 and 89

9.3 Character streams（字符流）

Java platform stores characters with Unicode convention. For most applications, I/O with character streams is no more complicated than I/O with byte streams. The character input and output stream classes automatically translate the Unicode characters to and from the local characters, usually an 8-bit superset of ASCII, without extra effort by the programmer.

For character sources and destinations, such as disk character files, you can simply use the character stream classes without paying much attention to character set issues. Figure 9-3 shows the hierarchical structure of the character stream classes.

Java 采用 Unicode 字符集。对于多数应用，字符 I/O 与字节 I/O 的处理方法相近。字符输入/输出流类会自动地将 Unicode 字符转换成本地设备所用的字符，通常是 ASCII 字符，无须额外的转换代码。

对于字符数据源和数据终端，如字符磁盘文件，直接使用字符流类，无须理会双方采用什么样的字符集。图 9-3 是字符流类的层次图。

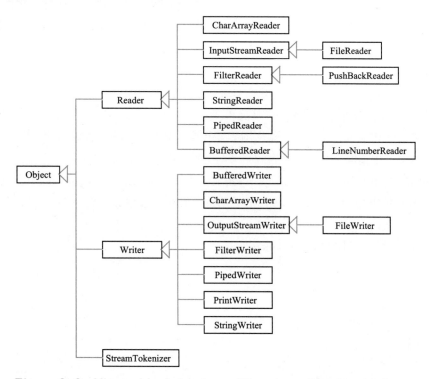

Figure 9-3 Hierarchical structure of the character stream classes

9.3.1　Text file vs Binary file（文本文件与二进制文件）

Data stored in a text file (character file) is represented in human-readable form. Data stored in a binary file is represented in binary form. You cannot read binary files. They are designed to be read by programs. For example, Java source programs are stored in text files and can be read by a text editor such as "notebook", but Java bytecode files are stored in binary files and are read by the JVM. The advantage of binary files is that they are more efficient to process than text files.

Text I/O requires encoding and decoding. When writing characters to a file, the stream classes convert Unicode characters to the file specific characters, that is, encoding. When reading characters from a file, the file specific characters are converted to Unicode characters, that is, decoding. Binary I/O does not require any conversions. When you write a byte to a file, the original byte is copied into the file. When you read a byte from a file, the exact byte in the file is returned.

Although it is not technically precise and correct, you can imagine that a text file consists of a sequence of characters and a binary file consists of a sequence of bits. For instance, the decimal integer 109 is stored as the sequence of three characters '1', '0', '9' in a text file and the same integer is stored as a byte value 6D in a binary file, because decimal 109 equals to hexadecinal 6D.

Figure 9-4 shows the difference between the storing of character and byte data.

文本（字符）文件的内容可读。二进制文件以二进制的形式存放，直接打开文件看到的是乱码，只能通过程序读取。例如，Java 源程序保存在文本文件里，可由像记事本这样的文本编辑程序打开并阅读。而 Java 的字节码文件是二进制文件，只能由 JVM 读取。二进制文件比文本文件容易处理。

文本文件的输入/输出需要编码和解码。向文件写入字符时，流类要把程序里的 Unicode 字符转换成文件规定的字符，即编码。从文件里读字符时，要把文件里的字符转换成 Unicode 字符，即解码。二进制文件的输入/输出不需要转换。将字节写入文件时，原始字节直接复制到文件中。从文件中读取字节时，读出的字节与文件里存放的形式一样。

粗略地说，文本文件由字符组成。二进制文件由一个个二进制位组成。以十进制整数 109 为例，在文本文件里是'1''0''9'三个字符，而在二进制文件里是 6D，109 的十六进制数是 6D。

图 9-4 显示了字符存储和字节存储的区别。

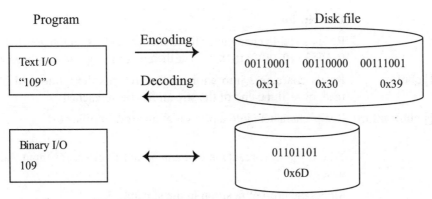

Figure 9-4　Difference between storing of character and byte data

9.3.2　Reader/Writer（Reader/Writer 类）

Character input streams are implemented by the subclasses of the Reader class and character output streams are implemented by the subclasses of the Writer class. The Reader class is an abstract class and the Writer class is an abstract class too. They are the super classes of all character stream classes, just like the InputStream and OutputStream are the super classes of all the other byte (binary) stream classes.

Since Java's native char is 16-bit Unicode, the Reader and Writer hierarchies support Unicode in all I/O operations.

字符输入由 Reader 类的子类实现，字符输出由 Writer 类的子类实现。Reader 和 Writer 都是抽象类，是所有其他字符流类的父类，就像 InputStream 和 OutputStream 是所有字节（二进制）流类的父类。

Java 采用 16 位的 Unicode 字符集，因此 Reader 和 Writer 都支持 Unicode 字符的读 / 写。

1. Reader

The Reader class is the super class of all classes of character input stream. All subclasses implement read(char[] cbuf, int off, int len) and close() methods of the Reader class. Most subclasses, however, will override some of the methods defined in the Reader class in order to provide higher efficiency, additional functionality, or both. Table 9-6 shows the common methods of Reader class.

Reader 是所有字符输入流类的父类，其子类都实现了 Reader 类的 read(char[] cbuf, int off, int len) 和 close() 方法。事实上，为了高效，增加新功能，多数子类都重写了父类的一些方法。Reader 类的常用方法如表 9-6 所示。

Table 9-6　Common methods of Reader class

Methods	Description
read()	Reads a single character. The character read, as an integer in the range 0 to 65535 (0x00-0xffff), or -1 if the end of the stream has been reached
read(char[] cbuf)	Reads characters into an array and returns the number of characters read, or -1 if the end of the stream has been reached
read(char[] cbuf, int off, int len)	Reads characters into a portion of an array at offset off
ready()	True if the next read() is guaranteed not to block for input, false otherwise
mark()	Marks the present position in the stream
reset()	Resets the stream. If the stream has been marked, attempt to reposition it at the mark
close()	Closes the stream and releases any system resources associated with it

2. Writer

The Writer class is the super class of all classes representing an output stream of characters. The subclasses must implement writer(char[] cbuf, int off, int len), flush() and close() methods of the Writer class in order to provide higher efficiency, additional functionality, or both. Table 9-7 shows the common methods of Writer class.

Write 类是所有字符输出流类的父类。子类要实现 Write 类的 write(char[] cbuf, int off, int len)、flush() 和 close() 方法。事实上，为了实现高效，增加新功能，多数子类都重写了父类的方法。Writer 类的常见方法如表 9-7 所示。

Table 9-7　Common methods of Write class

Methods	Description
write(int c)	Writes a single character
write(char[] cbuf)	Writes an array of characters
write(char[] cbuf, int off, int len)	Writes a portion of an array of characters
write(String str)	Writes a string
write(String str, int off, int len)	Writes a portion of a string
append(char c)	Appends the specified character to this writer
flush()	Flushes the stream
close()	Closes the stream, flushing it first

9.3.3　InputStreamReader/OutputStreamWriter（InputStreamReader/OutputStreamWriter 类）

There are times when you need to use classes from the "byte" hierarchy in combination with classes in the "character" hierarchy. To accomplish this, there are "adapter" classes:

InputStreamReader and OutputStreamWriter. InputStreamReader converts InputStream to Reader, and OutputStreamWriter converts OutputStream to Writer.

有时，既要使用字节流的类又要使用字符流的类，为此，Java I/O 类还提供 InputStreamReader 和 OutputStreamWriter 类作为"桥"。InputStreamReader 将 InputStream 转换成 Reader，OutputStreamWriter 将 OutputStream 转换成 Writer。

1. InputStreamReader

An InputStreamReader is a bridge from byte streams to character streams. It reads bytes and decodes them into characters using a specified charset. The charset may be specified by name explicitly, or takes the platform's default charset. Table 9-8 shows the common methods of InputStreamReader class. The prototypes of InputStreamReader's constructors are as below.

InputStreamReader 是字节流到字符流的桥，它读取字节，并按照指定的字符集将其解码为字符。可以由名字指定字符集或者采用平台的字符集。InputStreamReader 类的常见方法如表 9-8 所示。InputStreamReader 构造方法的原型如下。

```
public class InputStreamReaderextends Reader {
  InputStreamReader(InputStream in) {
    //Creates an InputStreamReader object, uses the default charset.
    …
  }
  InputStreamReader(InputStream in, String charsetName) {
    //Creates an InputStreamReader object, uses the named charset.
    …
  }
  InputStreamReader(InputStream in, Charset cs) {
    //Creates an InputStreamReader object, uses the given charset.
    …
  }
  InputStreamReader(InputStream in, CharsetDecoder dec) {
    //Creates an InputStreamReader object, uses the given charset decoder.
    …
  }
  …
}
```

Table 9-8　Common methods of InputStreamReader class

Method	Description
close()	Closes the stream and releases any system resources associated with it
read()	Reads a single character
read(char[] cbuf, int offset, int length)	Reads characters into a portion of an array
ready()	Tells whether this stream is ready to be read

【Example 9-10】Count characters and capital letters

```java
import java.io.*;
class CountCap {
public static void main(String[] args) throws IOException {
  int ch, total, capital=0;
  FileInputStream in;
  in=new FileInputStream("f:/Hello.java");
  InputStreamReader isr=new InputStreamReader(in);
  for(total=0;(ch=isr.read())!=-1;total++) {
    if(ch>='A'&&ch<='Z')
      capital++;
  }
  isr.close();
  in.close();
  System.out.println(total+"chars, "+capital+"capitals");
  }
}
```

 Output:

348chars, 16capitals

2. OutputStreamWriter

OutputStreamWriter is a bridge from character streams to byte streams. Characters written to it are encoded into bytes using a specified charset. The charset may be specified by name or take the platform's default charset.

Each invocation of a write() method causes the encoding converter to be invoked on the given characters. The resulting bytes are accumulated in a buffer before being written to the underlying output stream. Note that the characters passed to the write() method are not buffered. Table 9-9 shows the common methods of OutputStreamWriter class. The prototypes of OutputStreamWriter's constructors are as below.

OutputStreamWriter 是字符流到字节流的桥。它将写入的特定字符集的字符编码成字节，可以由名字指定字符集，或者采用平台的字符集。

调用 write() 方法会触发编码转换器对指定字符集的字符进行转换。输出之前，转换后的字节被放入缓冲区。注意，write() 的字符不缓冲。OutputStreamWriter 类的常见方法如表 9-9 所示。OutputStreamWriter 构造方法的原型如下。

```java
public class OutputStreamWriter extends Writer {
  OutputStreamWriter(OutputStream out) {
    //Creates an OutputStreamWriter object, uses the default character encoding.
    …
  }
  OutputStreamWriter(OutputStream out, String charsetName) {
    //Creates an OutputStreamWriter object, uses the named charset.
  }
  OutputStreamWriter(OutputStream out, Charset cs) {
```

```
    //Creates an OutputStreamWriter object, uses the given charset.
    …
  }
  OutputStreamWriter(OutputStream out, CharsetEncoder enc) {
    //Creates an OutputStreamWriter object, uses the given charset encoder.
    …
  }
  …
}
```

Table 9-9　Common methods of OutputStreamWriter class

Methods	Description
getEncoding()	Returns the name of the character encoding being used by this stream
flush()	Flushes the stream
write(int c)	Writes a single character
write(char[] cbuf, int off, int len)	Writes a portion of an array of characters
close()	Closes the stream, flushing it first
write(String str, int off, int len)	Writes a portion of a string

【Example 9-11】Copy a file

```
import java.io.*;
public class ReadBin {
  public static void main(String args []) throws IOException {
    int ch, total, capital=0;
    FileInputStream in;
    in=new FileInputStream("f:\\Hello.class");
    FileOutputStream out;
    out=new FileOutputStream("f:\\aaa\\Hello.class");
    InputStreamReader isr=new InputStreamReader(in);
    OutputStreamWriter osw=new OutputStreamWriter(out);
    while((ch=isr.read())!=-1) {
      osw.write(ch);
    }
    isr.close();
    in.close();
    osw.close();
    out.close();
  }
}
```

9.3.4　FileReader/FileWriter（FileReader/FileWriter 类）

The two classes support reading or writing characters from and to a file using default character encoding.

FileReade 类和 FileWriter 类支持字符文件的读 / 写，并按照默认方式对字符编码 / 解码。

1. FileReader

FileReader is meant for reading a stream of characters from a file. The constructors of the class assume that the default character decoding and the default buffer size are appropriate. It is possible to specify the decoding mode and the buffer size by yourself. All methods of FileReader class are inherited from its super class Reader. The prototypes of the constructors are as below.

FileReader 从文件读取字符流，类的构造方法采用默认的字符解码方式和默认的缓冲区大小，也可以自行设定解码方式和缓冲区大小。FileReader 类的方法都来自它的父类 Reader。FileReader 类构造方法的原型如下。

```
public class FileReader extends InputStreamReader {
  public FileReader(File file) throws FileNotFoundException {
    //Creates a new FileReader object and read from file "file".
    …
  }
  public FileReader(FileDescriptor fd) {
    //Creates a new FileReader object and read from FileDescriptor fd
    …
  }
  public FileReader(String fileName) throws FileNotFoundException {
    //Creates a new FileReader object and read from file fileName.
    …
  }
  …
}
```

Example 9-12 reads a group of character from a file. The "read" method of the Reader class is overridden in the FileReader class. The "read" method reads an array of characters from the file and returns the number of characters read. The array size in Example 9-12 is smaller than the number of chracters in the file, so the "read" method is invoked more than one time.

例 9-12 从文件里读字符。FileReader 类重写了 Reader 类的 read 方法。Read 方法从文件读取数组所能容纳的字符并返回所读字符个数。例 9-12 的数组不足以容纳文件里的所有字符，因此不止一次地调用了 read 方法。

【Example 9-12】**Read characters from a file**

```
import java.io.*;
class ReadChar {
  public static void main(String[] args) {
    char[] str=new char[100];
    int ch;
    try {
      FileReader in=new FileReader("f:\\hello.txt");
      While((ch=in.read(str))>0) {
        System.out.print(new String(str,0,ch));
```

```
            //System.out.print(str);
            //output all characters (some are invisible) of the array
        }
        in.close();
    }
    catch(IOException iox) {System.out.println("File not exist!"); }
  }
}
```

Output:

FileReader is meant for reading a stream of characters from a file.
It is possible to specify the encoding mode and the buffer size by yourself.

2. FileWriter

FileWriter is meant for writing streams of characters. The constructors of the class assume that the default character encoding and the default buffer size are acceptable. It is possible to specify the encoding mode and the buffer size by yourself. All methods of FileWriter class are inherited from its super class Writer. The prototypes of the constructors are as below.

FileWriter 向文件里写字符。FileWriter 类的构造方法采用默认字符编码方式和默认缓冲区大小，也可以自行设定编码方式和缓冲区大小。FileWriter 类的所有成员方法都来自它的父类 Writer。FileWriter 类构造方法的原型如下。

```
public class FileWrite rextends OutputStreamWriter {
   public FileWriter(File file) throws IOException {
      …
   }
   public FileWriter(File file) throws IOException {
      …
   }
   public FileWriter(File file, boolean append) throws IOException) {
      //Constructs a FileWriter object. If the second argument is true,
append characters to the end of the file.
      …
   }
   …
}
```

Example 9-13 and 9-14 write lines of characters and append lines of characters to file abc.txt respectively.

【Example 9-13】 **Create abc.txt and write several lines to it**

```
import java.io.*;
public class WriteChar {
   public static void main(String args []) throws IOException {
      String s1="Hello, every body";
      String s2="\r\nToday,we will study the java progrmming language!";
      //return and new line
```

```
        FileWriter fw=new FileWriter("f:\\abc.txt");
        fw.write(s1);
        fw.write(s2);
        fw.close();
    }
}
```

【Example 9-14】Append more characters

```
import java.io.*;
public class AppendTxt {
  public static void main(String[] args) {
    String fileName="f:\\Hello.txt" ;
    try {
      FileWriter writer=new FileWriter(fileName, true);
      writer.write("Add more.");
      writer.write("This is appended contents.\r\n" );
      writer.write("The file is different.\r\n");
      writer.write(" 输出一行中文也可以 .\r\n");
      writer.close();
    }
    catch (IOException iox) {
      System.out.println("Problem writing"+fileName);
    }
  }
}
```

9.3.5 BufferReader/BufferWriter（BufferReader/BufferWriter 类）

The buffered read and write can speed up the file reading and writing. BufferedReader and BufferedReader provide buffers while reading or writing.

1. BufferReader

BufferedReader reads text from a character input stream, buffering characters allows efficient reading of characters, arrays and lines. The buffer size may be specified, or take the default that is large enough for most purposes.

Without buffering, each invocation of read() could cause bytes to be read from the file, converted into characters and then returned, which can be very inefficient. Wrap read operation with BufferedReader is advisable, for instance.

```
BufferedReader in = new BufferedReader(new FileReader("foo.in"));
```

The prototypes of the constructors are as below.

BufferedReader 从字符输入流中读取文本，缓冲机制能够实现字符、数组和一行字符的高效读取。可以自行设定缓冲区的大小，也可以采用默认缓冲区大小。多数情况下，默认缓冲区足够大。

如果没有缓冲，每次 read() 调用或都会导致从文件中读取字节，转换成字符再返回的低效处理。建议用 BufferedReader 处理读取操作，如：

```
BufferedReader in = new BufferedReader(new FileReader("foo.in"));
```
BufferedReader 类构造方法的原型如下。
```
public class BufferedReaderextends Reader {
  public BufferedReader(Reader in) {
  //Creates a buffering character-input stream that uses the default buffer size.
    …
  }
  public BufferedReader(Reader in, int sz) {
  //Creates a buffering character-input stream that uses the specified buffer size.
    …
  }
    …
}
```

BufferedReader class adds an important method readLine(). The method reads a line of text. A line is considered to be terminated by any one of a line feed '\n', a carriage return '\r' or both of them "\r\n". The readLine() returns the text of the line, but no includes any terminal characters. When readLine() returns null, you're at the end of the file.

BufferedReader 类增加了 readLine() 方法。该方法每次读取一行文本，文件里的一行以换行 '\n'、回车 '\r' 或回车加换行结尾。方法返回包含该行内容的字符串，不包含任何行终止符，如果已到达末尾，则返回 null。

【Example 9-15】Method readLine() reads a line of text

```
import java.io.*;
public class ReadTest {
  public static void main(String args []) {
    try {
      String str;
      FileReader reader=new FileReader("f:\\abc.txt");
      BufferedReader br=new BufferedReader(reader);
      while((str=br.readLine())!=null) {
        System.out.println(str);
      }
      br.close();
      reader.close();
    }
    catch(IOException e) {
      e.printStackTrace();
    }
  }
}
```

2. BufferedWriter

BufferedWriter writes text to a character output stream, buffering characters speeds up the writing operation, provides efficient writing of single characters, arrays and strings.

The buffer size may be specified or be the default. The default is large enough for most purposes.

BufferedWriter 将文本写入字符输出流，缓冲机制提高了写入速度，可以高效输出字符、数组和字符串。

可以指定缓冲区的大小，或者采用默认大小。在大多数情况下，默认大小足够。

```
public class BufferedWriter extends Writer {
  public BufferedWriter(Writer out) {
     //Creates a buffered character-output stream and uses the default buffer sized.
     …
  }
  public BufferedWriter(Writer out, int sz) {
     //Creates a new buffered character-output stream and uses the given buffer size sz.
     …
  }
     …
}
```

A newLine() method is added to the BufferedWriter class, which uses the platform's notion of line separator. Not all platforms use the newline character ('\n') to terminate lines. Calling this method to terminate each output line is therefore preferred to writing a newline character directly. Example 9-16 read and write lines of text with readline() and newline().

BufferedWriter 类增加了 newLine() 方法，用于写入行分隔符，并非所有平台都使用换行符 '\n' 做行的结束符。因此该方法优于直接写入新行符 "\r\n"。

【**Example 9-16**】**Read from a file and write to another file with the readLine() method**

```
import java.io.*; public class CopyTxt{
public class CopyTxt {
  public static void main(String[] args) {
    String fileName1="f:\\Hello.txt" ;
    String fileName2="e:\\Hello.txt" ;
    String str;
    try {
      FileReader reader=new FileReader(fileName1);
      FileWriter writer=new FileWriter(fileName2, true);
      BufferedReader br=new BufferedReader(reader);
      BufferedWriter bw=new BufferedWriter(writer);
      while((str=br.readLine())!=null) {
        bw.write(str);
        bw.newLine(); //these two can be replaced with bw.write(str+"\r\n");
      }
      br.close();
      bw.close();
    }
    catch (IOException e) {
      e.printStackTrace();
    }
  }
}
```

9.3.6 PrintWriter（PrintWriter 类）

PrintWriter prints formatted representations of objects to a text output stream. This class implements all of the print methods in PrintStream.

Unlike the PrintStream class, if automatic flushing is enabled it will be done only when one of the println, printf, or format methods is invoked, rather than whenever a newline character happens to be output. These methods use the platform's own notion of line separator rather than the newline character. Prototype of the constructors of the PrintWriter class are as below.

PrintWriter 向文本输出流输出对象的格式化处理结果。该类实现了 PrintStream 的所有 print 方法。与 PrintStream 类不同，只有调用了 println、printf 或 format 方法时才自动刷新而不是在输出换行符时刷新。这些方法使用平台自己的行结束符而不用 newline 指定的行结束符。PrintWriter 类构造方法的原型如下。

```
public class PrintStream extends FilterOutputStream implements
Appendable, Closeable {
    public PrintWriter(File file)throws FileNotFoundException {
      //Creates a new PrintWriter, without automatic line flushing.
      …
    }
    public PrintWriter(String fileName)throws FileNotFoundException {
      //Creates a new PrintWriter object, without automatic line flushing.
      …
    }
    public PrintWriter(OutputStream out) {
      //Creates a new PrintWriter object, without automatic line flushing.
    }
    public PrintWriter(Writer out) {
      //Creates a new PrintWriter object, without automatic line flushing.
    }
    …
}
```

9.4 File class（文件类）

There is a File class in java.io package, which provides several sophisticated file dealing methods. The "File" can represent either the name of a particular file or the names of a set of files in a directory. If it's a set of files, you can look at that set using the list() method, which returns an array of String. It makes sense to return an array rather than one of the flexible container classes, because the number of elements is fixed, and if you want a different directory listing, you just create a different File object. This section shows an example of the use of the File class, including the associated FilenameFilter interface.

The File class is more than just a representation for an existing file or directory. You can also use a File object to create a new directory or an entire directory path if it doesn't exist. You can also look at the characteristics of files (size, last modification date, read/write), judge whether a File object represents a file or a directory and delete a file. The prototypes of the constructors of the File class are as below.

在 java.io package 包里有个 File 类，它提供了一些非常成熟的文件处理方法。File 既可以代表一个特定文件的名，也可以代表某个目录里一系列文件的名字。如果代表一个文件集合，可用 list() 方法查看目录里的文件，list() 方法返回一个 String 数组。之所以要返回一个数组，而非某个集合，是因为文件的数目是固定的。若想得到其他目录的文件列表，只需创建一个不同的 File 对象。本节通过一个例子介绍 File，包括相关的 FilenameFilter（文件名过滤器）接口的使用。

File 类不仅表示路径或者文件，如果目录不存在亦可用一个 File 对象新建一个目录，甚至一个完整的目录路径。也可以通过 File 对象获取文件的属性（大小、上一次的修改日期、读/写属性等），判断 File 对象代表文件还是目录，以及删除文件等。File 类构造方法的原型如下。

```
public class File extends Object implements Serializable, Comparable<File> {
    public File(String pathname) {
       //Creates a new File object by converting the given pathname string into an abstract pathname.
       …
    }
    public File(String parent, String child) {
       //Creates a new File object from a parent pathname string and a child pathname string.
       …
    }
    public File(File parent, String child) {
       //Creates a new File object from a parent pathname string and a child pathname string.
       …
    }
    …
}
```

When an object of the File class is created, if the file does not exist, a file will not be created automatically. You must call the createNewFile() method to create it. Table 9-10 shows the common methods of File class.

当创建一个 File 类的对象时，如果对应的文件不存在，系统不会自动创建文件，必须调用 createNewFile() 方法来创建文件。表 9-10 是 File 类的一些常用方法。

Table 9-10 Common methods of File class

Method	Description
canRead()	Tests whether the application can read the file denoted by the abstract pathname
canWrite()	Tests whether the application can modify the file denoted by the abstract pathname
exists()	Tests whether the file or directory denoted by the abstract pathname exists.
isFile()	Tests whether the file denoted by the abstract pathname is a normal file
isHidden()	Tests whether the file named by this abstract pathname is a hidden file
length()	Returns the length of the file denoted by this abstract pathname
createNewFile()	Atomically creates a new, empty file named by this abstract pathname if and only if a file with this name does not yet exist
delete()	Deletes the file or directory denoted by this abstract pathname
mkdir()	Creates the directory named by this abstract pathname
getName()	Returns the name of the file or directory denoted by this abstract pathname
getPath()	Converts this abstract pathname into a pathname string
isAbsolute()	Tests whether this abstract pathname is absolute
getAbsolutePath()	Returns the absolute pathname string of this abstract pathname

【Example 9-17】Display information about the file

```
import java.io.*;
public class ReadFile {
  public static void main(String args []) throws IOException {
    File file=new File("f:\\abc.txt");
    System.out.println("file name:"+file.getName());
    System.out.println("file exist:"+file.exists());
    if(file.exists()==false) {
      file.createNewFile();
      System.out.println("file exist:"+file.exists());
    }
    System.out.println("relative path:"+file.getPath());
    System.out.println("absolut path:"+file.getAbsolutePath());
    System.out.println("readable:"+file.canRead());
    System.out.println("writeable:"+file.canWrite());
    System.out.println("file size:"+file.length());
  }
}
```

 Output:

file name:abc.txt

file exist:true

relative path:f:\abc.txt

absolut path:f:\abc.txt
readable:true
writeable:true
file size:68

Exercises

1. Write a program, input several characters from the keyboard and print them.

(1) Use InputStream object to connect the keyboard.

(2) Use OutputStream object to connect the console.

写程序，从键盘输入几个字符，再输出。

(1) 利用 InputStreaml 对象与键盘关联。

(2) 利用 OutputStream 对象与显示器关联。

2. Write a program to write 10 numbers to file f:\aaa.dat.

(1) Define a byte array and assign 10 numbers to the array.

(2) Create FileOutputStream object and write the array contents to the file.

写程序，将 10 个数字写入 f:\aaa.dat 文件中。

(1) 定义一个字节数组，为数组赋值。

(2) 创建一个 FileOutputStream 类的对象，将数组里的 10 个数写入文件。

3. Write a program, append 10 numbers to file f:\aaa.dat.

(1) Define a byte array, and assign 10 numbers to the array.

(2) Create FileOutputStream object and append the array contents to the file.

写程序，将 10 个数字追加到 f:\aaa.dat 文件。

(1) 定义一个字节数组，为数组赋值。

(2) 创建 FileOutputStream 类的对象，将数组里的 10 个数追加到 f:\aaa.dat 文件。

4. Write a program, read from file f:\aaa.dat. Create a FileInputStream object, read the file and display the values read.

写程序，读取 f:\aaa.dat 文件的内容。创建 FileInputStream 类的对象，读取文件，输出读到的内容。

5. Write a program, write several strings to file f:\aaa.txt.

(1) Create a FileWriter object and write strings to the file.

(2) Open the file in a text reader to see what you've written in the file.

写程序，向文件 f:\aaa.txt 里写入几个字符串。

(1) 创建一个 FileWriter 类的对象，向文件里写入几个字符串。

(2) 在文本编辑软件下打开文件，看看你写入的内容。

6. Write a program, append more strings to file f:\aaa.txt.

(1) Create a FileWriter object and append strings to the file.

(2) Open the file in a text reader to see what you've written in the file.

写程序，追加更多的字符串到 f:\aaa.txt 文件。

(1) 创建 FileWriter 类的对象，往文件里写几个字符串。

(2) 在文本编辑软件下打开文件，看看你写入的内容。

7. Write a program, read from the file f:\aaa.txt.

Create a FileReader object, read the file and print what you read.

写程序，读取 f:\aaa.txt 并输出文件的内容。

创建 FileReader 类的对象，读取文本文件的内容，并将内容输出到显示器上。

8. Write a program, read lines from the keyboard and write it to file f:\kk.txt.

(1) Create a InputStreamReader object.

(2) Create a BufferedReader object, and invoke readLine method to read strings

(3) Create a FileWriter object to write strings to the file.

写程序，从键盘读取一整行字符并写到 f:\kk.txt 文件。

(1) 创建 InputStreamReader 类的对象。

(2) 创建 BufferedReader 类的对象，调用 readLine() 方法读取一行字符。

(3) 创建 FileWriter 类的对象，把字符串写到文件。

Chapter 10
Multi-threading（多线程）

10.1 Concept of multi-threading（多线程的概念）

10.1.1 What's a thread?（什么是线程？）

Let's start from multi-tasking, a computer allows more than one program running at the same time means that multi tasks are carrying out. If we extend the idea of multi-tasking by allowing an individual program to execute multiple tasks, the multi threads come out.

Each task within the program is called a thread. A thread is a single sequential flow of control within a process.

Concurrent programming allows you to divide a program into separate, independently running tasks.

10.1.2 Thread vs process（线程与进程）

A process is a self-contained running program and can be considered an application, whereas a thread is a task in the process. Threads are subsets of a process, and a process can have many threads, that executing different tasks in parallel. Each process has its own resources, such as memory, and all threads in a process share the same resources of the process.

一个进程是一个独立的运行程序，可以被看成是一个应用，而一个线程是进程的一个任务。线程是进程的子集，一个进程可以有很多线程，多个线程并行执行不同的任务。每个进程拥有自己的资源，例如存储空间，而进程内的线程共享这个进程的资源。

Within a process, control usually follows a single thread, typically starting with the first statement of the main, stepping through a sequence of statements, and ending. This is the single thread programming, as shown in Figure 10-1.

Java supports multiple concurrent threads' execution or multi-threading. A two threads program that performs more than one task is as shown in Figure 10-2.

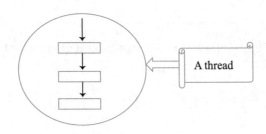

Figure 10-1 Single thread

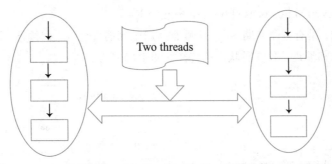

Figure 10-2 Multi-threading

Java supports concurrent programming via threads. A thread is a single sequential flow of execution within a process.

A multi-tasking OS allows running more than one process at a time. A process can have multiple concurrently executing threads. For a multiple CPUs system, multiple threads can run simultaneously. A multiple CPUs and multiple threads system is as shown in Figure 10-3. For a single CPU system, the multiple threads run concurrently following special rules. A single CPU, multi-threads system is as shown in Figure 10-4.

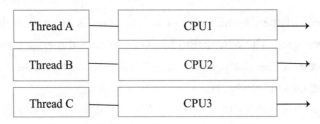

Figure 10-3 Multi threads and CPUs

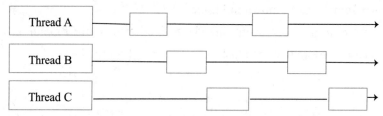

Figure 10-4　Multi-threads and single CPU

10.2　Life cycle of a thread（线程的生命周期）

The status of a thread is dynamically changed. From its born to dead, a thread has a life cycle that includes seven states: newborn, ready, run, wait, sleep, blocked and dead. The transition of the seven states is as shown in Figure 10-5.

线程是动态的，是有生命周期的。一个线程从新生到消亡包含 7 种状态：新建、就绪、运行、等待、休眠、阻塞和消亡。这 7 种状态的转换如图 10-5 所示。

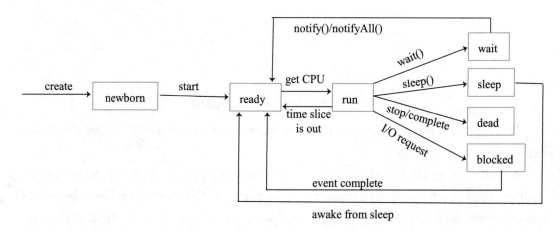

Figure 10-5　Life cycle of a thread

1. Newborn

When an object of the Thread class or its subclass is created, a new thread is born, which is in newborn state. The new thread has been allocated memory and its data has been initialized, but the thread has not yet been scheduled. The start() method of the Thread class can make the thread into the ready state.

创建一个 Thread 类或其子类的对象时，一个新的线程就诞生了并处于新建（newborn）状态。线程已被分配存储空间，数据已被初始化，但该线程还未被调度。Thread 类的 start() 方法会让线程进入就绪状态。

2. Ready

Invoking start() method of a thread object makes a thread from the newborn state enter into the ready state. The thread in this state is ready to run. Each thread has its own run() method which is alike the main method, contains a sequential statements. The run() method executes (the thread starts to run) as long as the thread gets the CPU.

调用线程对象的 start() 方法，可让线程从新生状态转入就绪状态，处于就绪状态的线程具备了运行条件。每个线程都有自己的 run() 方法，它类似 main 方法，包含顺序执行的语句。只要获得 CPU，线程的 run() 方法就执行（线程开始运行）。

3. Run

Threads in the run state may enter into other states depending on the following situations:

(1) The allocated time slice is used up and the thread has not yet finished executing, the thread enters into the ready state and waits for the CPU to schedule again.

(2) The call of its wait() method makes the thread go into the wait state until another thread wakes it up by invoking the notify() or notifyAll() method.

(3) The call of its sleep(time-delay) method forces the thread to go into the sleep state. As the time-delay milliseconds passes away, the thread will automatically back to the ready state.

(4) Execution is finished or is forced to terminate within a time slice, the thread enters into the dead state.

(5) An I/O request is encountered which forces the thread to enter into the blocked state.

处于运行状态的线程会在以下情形转入其他状态：

（1）一个时间片用完，线程的执行尚未结束，线程进入就绪状态，等待 CPU 的再次调度。

（2）调用了 wait() 方法，进入等待状态，直到其他线程用 notify() 方法或 notifyAll() 方法将该线程唤醒。

（3）调用 sleep(time-delay) 方法，使线程进入休眠状态，休眠 time-delay 毫秒后，线程自动进入就绪状态。

（4）在一个时间片内线程执行完毕或被强行终止，进入消亡状态。

（5）在执行过程中遇到 I/O 请求，进入阻塞状态。

4. Wait

The thread in the running state will enter into the waiting state if the wait() method is invoked, meanwhile, the thread gives up the resources it owns. The thread keeps waiting until another thread calls the notify() or notifyAll() method to wake it up and makes the thread enter into the ready state for the CPU's next schedule.

调用 wait() 方法，使处于运行状态的线程让出所占用的资源并进入等待状态，直到其他线程调用 notify() 或 notifyAll() 方法将其唤醒，线程回到就绪状态排队，等待 CPU 的下一次调度。

5. Sleep

A running thread goes into sleep state if sleep() method is invoked. The resource is handed over. When the sleep time passes away, the thread goes to the ready state automatically and waits for the CPU's next schedule.

调用 sleep() 方法会使处于运行中的线程让出所占资源并进入休眠状态。休眠时间结束，线程会自动进入就绪状态，等待 CPU 的下一次调度。

6. Blocked

Due to fail to applying for resources or some events happen, such as I/O requests, a thread is forced to interrupt its normal executing and enters into the blocked state, until the required resources are available or the I/O request is completed, the thread in the blocked state goes into the ready state, waiting for the CPU's next schedule.

由于申请资源未得到满足或者某事发生，例如 I/O 请求，线程被迫中断运行并转入阻塞状态，直到线程获得了所需的资源或者完成了 I/O 任务，才会从阻塞状态回到就绪状态，等待 CPU 的再次调度。

7. Dead

If a thread finishes its execution within the allocated time slice or its execution is stopped for some reason, the thread goes to the dead state and ends its life cycle.

如果线程在分配的时间片内结束任务或者由于某种原因停止执行，线程进入消亡状态，结束其生命周期。

10.3 Creating threads（创建线程）

There are two approaches to create threads in Java. We mentioned that actually a thread is an object of Thread's subclass, so producing a thread is creating an object. The first is the direct approach, to create a thread by directly creating an object of the subclass of java.lang.Thread. The second is the indirect approach, to create an object needing to implement the java.lang.Runnable interface first.

Java 有两种方式创建线程。我们说过，线程事实上是 Thread 类的子类的对象，因此生成一个线程就是创建一个对象。方法一是直接法，直接创建 Thread 类的子类的对象。方法二是间接法，创建对象前先实现 java.lang.Runnable 接口。

10.3.1 Direct approach of creating a thread（直接法创建线程）

The Thread class is in the java.lang package. It provides many methods for creating, manipulating and processing threads. Three constructors of the Thread class are as below:

```
public Thread(String threadName) {…}
public Thread() {…}
public Thread(Runnable target) {…}
```

The first constructor is used to create a thread object with the name threadName. The second one needs not to provide a thread name. The third one is used for indirectly creating a thread.

Thread 类在 java.lang 包中，该类定义了许多创建、操作和处理线程的方法。Thread 类有 3 个构造方法。

用第一个构造方法，创建名为 threadName 的线程。用第二个构造方法无须给出线程的名字。第三个构造方法在间接法创建线程时用。

The syntax for creating a new thread by inheriting the Thread class is below.

```
class CreateThread extends Thread {
  public void run() {
    …
  }
  public static void main(String args[]) {
    CreateThread thread=new CreateThread();
    thread.start(); //for calling the run method.
  }
}
```

The essential task of inheriting the Thread class is to override the run() method of the Thread class in the subclass. Each thread has its own run() method, which contains the codes that a thread needs to execute for its special purpose. Threads must be created and started in the main() method. After a thread is created, the start() method must be called to run the thread, otherwise the thread can't run. The start() method automatically calls the corresponding run() method to execute the thread when everything is ready. Example 10-1 is an example of computing factorials with a thread.

继承 Thread 类的重要任务是在子类里重写 Thread 类的 run() 方法。每个线程都有自己的 run() 方法，它包含了执行特定任务的代码。线程的创建与启动必须在 main() 方法中进行。线程生成后，必须调用 start() 方法启动，否则线程不会运行。当一切就绪，start() 方法会自动调用线程对应的 run() 方法运行线程。例 10-1 是利用线程计算阶乘的例子。

【Example 10-1】**Create a thread to compute 1!+2!+3!+4!+5!**

```
public class T3 extends Thread {
  public void run() {
    int s=0,t=1;
    for(int i=1;i<=5;i++) {
      t*=i;
      s+=t;
    }
    System.out.println("1!+2!+3!+4!+5!="+s);
  }
  public static void main(String[] args) {
    T3 th=new T3();
    th.start();
  }
}
```

 Output:

1!+2!+3!+4!+5!=153

10.3.2 Indirect approach of creating a thread（间接法创建线程）

Since Java subclasses can only have one super class, the direct approach of creating a thread makes it impossible for the thread to inherit any other class. The indirect approach gives threads the opportunity to inherit one or more super classes. There is an interface Runnable is in java.lang package too and has only one method run().

```
public interface java.lang.Runnable {
  public void run(); //only one method
}
```

Use this interface and make use of the third constructor of the Thread class:

```
public Thread(Runnable target) {…}
```

a thread can be created indirectly.

由于 Java 子类只能有一个父类，直接法使线程无法再继承其他的类。间接法使线程有机会继承一到多个父类。Java.lang 包里有一个 Runable 接口，它只有一个 run() 方法。利用这个接口和 Thread 的第三个构造方法，可以间接地创建线程。

The syntax for creating a new thread by implementing the Runnable interface is as below.

```
class CreateThread implements Runnable {
  public void run() {
    …
  }
  public static void main(String args[]) {
    CreateThread th=new CreateThread();
    Thread tha=new Thread(th);    //create a thread
    tha.start();                  //execute run method
  }
}
```

Creating a thread by implementing the Runnable interface is divided into four steps:

(1) Define a class that implements the Runnable interface. It is essential to implement the run() method of the Runnable interface. Explicitly, a Runnable type reference can refer to an object of the class.

(2) In main(), create an object of the class defined in step 1.

(3) In main(), create an object of the class Thread (a new thread), the object created in step 2 as the actual parameter of the constructor.

(4) Start the thread by calling start() method.

That is why we call it indirect, in comparison with the direct approach that needs only two steps for creating and starting a thread, there are two more steps in the indirect approach. Example 10-2 is an example of creating a thread indirectly. Example 10-3 is an example that the class extends a class and implements the Runnable interface at the same time.

通过实现 Runnable 接口创建线程需要四步：

（1）定义一个实现 Runnable 接口的类，必须实现 Runnable 接口中的 run() 方法。显然，Runnable 类型的引用可以指向这个类的对象。

（2）在 main() 方法中，创建在第一步里定义的类的对象。

（3）在 main() 方法中创建一个 Thread 类的对象（即新的线程），用在第二步里创建的对象做构造方法的实参。

（4）调用 start() 方法启动线程。

这就是把它叫做间接法的原因，与只需要 2 步的直接法比较，间接法多了 2 步。例 10-2 是间接法创建线程的例子。例 10-3 是一个既用继承类又同时用实现接口的方法创建线程的例子。

【Example 10-2】**Compute factorials 1!+2!+3!+4!+5! , create a thread indirectly**

```java
public class T3 implements Runnable {
  public void run() {
    int s=0,t=1;
    for(int i=1;i<=5;i++) {
      t*=i;
      s+=t;
    }
    System.out.println("1!+2!+3!+4!+5!="+s);
  }
  public static void main(String[] args) {
    T3 th=new T3();
    Thread tha=new Thread(th);
    tha.start();
  }
}
```

Output:

1!+2!+3!+4!+5!=153

【Example 10-3】**Extend a class and implement Runnable**

```java
class Studt {
  String name;
  int id;
  public Studt(String n,int d) {
    name=n;
    id=d;
  }
}
public class ThreadDemo extends Studt implements Runnable {
  private int age;
  public void run() {
    try {
      for(int i=0;i<2;i++) {
        System.out.println("My name "+name+" My ID "+id+" My Age "+age);
        Thread.sleep(200);
```

```
        }
      }
      catch(InterruptedException k) {}
    }
    public ThreadDemo(String n,int i,int a) {
      super(n,i);
      age=a;
    }
    public static void main(String args[]) {
      ThreadDemo td1=new ThreadDemo("Jeff",1024,23);
      ThreadDemo td2=new ThreadDemo("Tim",1026,25);
      new Thread(td1).start();
      new Thread(td2).start();
    }
}
```

🧩 Output:

My name Jeff My ID 1024 My Age 23
My name Tim My ID 1026 My Age 25
My name Jeff My ID 1024 My Age 23
My name Tim My ID 1026 My Age 25

10.4 Main thread（主线程）

Each Java application program has a most important thread, named "main" and it is started automatically. The main thread executes the main method (a regular thread executes run method). All other threads must be created and started within the main method. Example 10-4 let main thread sleep for a while.

每个Java应用程序都有一个重要的线程，主线程，它是自动启动的。主线程执行main方法（普通线程执行run方法）。其他线程必须在main方法里创建与启动。例10-4让主程序休眠一会儿。

【Example 10-4】Let the main thread sleep for a while

```
public class TheMain {
  public static void main(String[] args) {
    try {Thread.sleep(2000); }
    catch(InterruptedException e){ }
    System.out.println("How are you?");
    TheMain b=new TheMain();
    b.fun();
  }
  void fun() {
    try {Thread.sleep(3000); }
    catch(InterruptedException e) {}
    System.out.println("Fine, thank you.");
```

}
}

 Output:

How are you?

Fine, thank you.

If several threads are in the runnable state, even the main thread comes to an end, the program won't end, until all threads end. Each thread's dead time depends on the competing result and its task. For threads starting at the same time, when or which of them runs first depends on who gets the CPU first. Example 10-5 is an example that the main thread ends running before the other threads.

Usually, the program runs with at least two threads in the runnable state (ready and run states). One is the main thread and the other is the garbage collection thread.

如果多个线程都处于 runnable 状态，即使主线程结束，程序也不会结束，直到所有线程结束运行。每个线程的消亡时间取决于多个线程的竞争结果以及具体任务的情况。对于同时启动的线程，至于什么时候，哪个先执行，取决于谁先获得 CPU。例 10-5 是主线程先于其他线程结束运行的例子。

通常，程序运行时至少 2 个线程处于可运行状态（就绪或者运行状态）。一个是 main 线程，另一个是垃圾收集线程。

【Example 10-5】**Main thread ends up before the other threads**

```java
class NewThread extends Thread {           //define a subclass of thread
  String name;
    NewThread(String tdname) {
    name=tdname;
    System.out.println(currentThread());   //return an thread object
  }
  public void run() {                      //override run
    try {
      for(int i=3;i>0;i--) {
        System.out.println(name+":"+i);
        sleep(200);
      }
    }
    catch(InterruptedException e) {
      System.out.println(name+"Interrupted.");
    }
    System.out.println(name+" Exiting.");
  }
}
class ThreadDemo {
  public static void main(String args[]) {
    NewThread one=new NewThread("One");
    NewThread two=new NewThread("Two");
```

```
    NewThread three=new NewThread("Three");
    one.start();
    two.start();
    three.start();
    try {
      Thread.sleep(200);
    }
    catch(InterruptedException e) {
      System.out.println("Main Thread Interrupt");
    }
    System.out.println("main Thread Exiting.");
  }
}
```

Output:

Thread[main,5,main]

Thread[main,5,main]

Thread[main,5,main]

One:3

Two:3

Three:3

Two:2

One:2

main Thread Exiting.

Three:2

Two:1

One:1

Three:1

Three Exiting.

Two Exiting.

One Exiting.

10.5 Methods of Thread class（线程类的方法）

Some methods of Thread class is listed in Table 10-1. Example 10-6 to 10-8 are examples of some methods calling for different purposes.

Table 10-1 Methods of Thread class

Method	Function
isAlive()	判断线程是否处于激活状态，即是否终止
currentThread()	返回处于 run 状态的线程对象

续表

Method	Function
getPriority()	获得线程的优先级
setPriority()	设置线程的优先级
sleep(int n)	让线程睡眠 n 毫秒
wait()	将当前对象进入等待状态
join()	加入新线程
yield()	放弃
notify()/notifyAll()	唤醒等待状态中的一个或所有等待线程

【Example 10-6】 sleep method, delay a second after output a message

```java
class My_Thread extends Thread {
  public void run() {
    try {
      for(int i=0;i<3;i++) {
        System.out.println("java study");
        sleep(1000);
      }
    }
    catch(InterruptedException e) {
      System.out.println(e.getMessage());
    }
  }
}
public class Thread2 {
  public static void main(String[] args) {
    My_Thread a=new My_Thread();
    a.start();
  }
}
```

Output:

java study

java study

java study

【Example 10-7】 join method

```java
public class TestThread4 {
  public static void main(String[] args) {
    MyThread2 thread2=new MyThread2("mythread");
    thread2.start();
    try {
      thread2.join();
    }
    catch(InterruptedException e) {
      e.printStackTrace();
    }
```

```
      for(int i=0;i<=3;i++) {
        System.out.println("I am main Thread");
      }
    }
}
class MyThread2 extends Thread {
  MyThread2(String s) {
    super(s);
  }
  public void run() {
    for(int i=1;i<=3;i++) {
      System.out.println("I am a\t"+getName());
      try {
        sleep(1000);
      }
      catch(InterruptedException e) {
        return;
      }
    }
  }
}
```

Output:

I am a   mythread
I am a   mythread
I am a   mythread
I am main Thread
I am main Thread
I am main Thread
I am main Thread

【Example 10-8】 **yield method**

```
public class TestThread5 {
  public static void main(String[] args) {
    MyThread3 t1=new MyThread3("t1");
    MyThread3 t2=new MyThread3("t2");
    t1.start();
    t2.start();
    for(int i=0;i<=5;i++)
    System.out.println("I am main Thread");
  }
}
class MyThread3 extends Thread {
  MyThread3(String s) {
    super(s);
  }
  public void run() {
    for(int i=1;i<=5;i++) {
```

```
      System.out.println(getName()+":"+i);
      if(i%2==0) {
        yield(); //give up
      }
    }
  }
}
```

Output:

I am main Thread
I am main Thread
I am main Thread
I am main Thread
I am main Thread
I am main Thread
t1:1
t2:1
t1:2
t1:3
t2:2
t2:3
t2:4
t1:4
t1:5
t2:5

The running result may be different each time.

10.6　Thread synchronization（线程同步）

Java supposes multi-thread concurrent control. When multiple threads running at the same time and sharing the same resources such as fields, might lead to inaccurate results. The confliction can be avoided by introducing in the synchronization locks, which guarantee that a thread's run will not be interrupted by another thread processing the same data.

Java 支持多线程的并发控制，当多个线程同时运行并共享资源，例如 fields，可能会导致数据不正确，这种冲突，可以通过引入同步锁来避免。同步锁能够保证线程的运行不会被其他处理同一数据的线程中断。

How to realize the synchronization of multi-threads?
如何实现多线程的同步呢？

1. Synchronizing methods with keyword "synchronized"

In fact, every object in Java has a built-in lock. Put keyword "synchronized" keyword in front of a method, then the method is protected. Before calling the protected method, the built-in lock needs to be got otherwise the thread calling the method will be blocked. For instance:

```
public synchronized void save() {}
```

实际上，Java 的每个对象都有一个内置锁，将 synchronized 放到方法前可对方法进行保护。调用该保护的方法前，需要获得内置锁，如果不能获得，调用该方法的线程就会进入阻塞状态。例如：public synchronized void save() { } 用于锁住 save() 方法。

2. Synchronizing blocks of codes with keyword "synchronized"

Keyword "synchronized" can be used to protect a block of statements. For instance:

```
synchronized (object) { }
```

The above approach to synchronization is a resource consuming solution, should be minimized. It is often not necessary to synchronize the entire method, just synchronize a block of important codes. Example 10-8 is an example of multi-threads' synchronization.

synchronized 关键字可以保护一段代码（一个语句块），用于保护语句块，例如：

```
synchronized (object) { }
```

上述实现同步的方法是一种消耗资源的解决办法，因此应尽量减少同步的范围。通常没有必要同步整个方法，仅仅对重要的代码进行同步即可。例 10-8 是一个多线程同步的例子。

【Example 10-9】**Synchronization of two threads**

```
public class SynchronizedThread {
  class Bank {
    private int account=100;
    public int getAccount() {
      return account;
    }
    public synchronized void save(int money) { //synchronize a method
      account+=money;
    }
    public void save1(int money) {
      synchronized (this) { //synchronize a block of codes
        account+=money;
      }
    }
  }
  class NewThread implements Runnable {
    private Bank bank;
    public NewThread(Bank bank) {
      this.bank=bank;
    }
    public void run() {
```

```
      for (int i=0; i<10; i++) {
        bank.save(10);
        System.out.println(i+"Account balance:"+bank.getAccount());
      }
    }
  }
  public void useThread() { //create thread, invoke inner class
    Bank bank=new Bank();
    NewThread new_thread=new NewThread(bank);
    System.out.println("thread1");
    Thread thread1=new Thread(new_thread);
    thread1.start();
    System.out.println("thread2");
    Thread thread2=new Thread(new_thread);
    thread2.start();
  }
  public static void main(String[] args) {
    SynchronizedThread st=new SynchronizedThread();
      st.useThread();
    }
  }
```

Output:

thread1

thread2

0Account balance:110

1Account balance:120

2Account balance:130

3Account balance:140

4Account balance:150

5Account balance:160

6Account balance:170

7Account balance:190

0Account balance:190

1Account balance:210

2Account balance:220

3Account balance:230

4Account balance:240

8Account balance:200

5Account balance:250

6Account balance:270

7Account balance:280

8Account balance:290
9Account balance:300
9Account balance:260

10.7 Communication between threads（线程间的通信）

Threads communicating with each other is realized by the wait(), notify() and notifyAll() methods. These methods can only be used in synchronized blocks or synchronized methods. The wait() method makes a thread enter into the wait state and release the lock simultaneously. Sleep() makes a thread enter into the sleep state but does not release the lock.

线程之间的通信是通过 wait()、notify() 和 notifyAll() 方法实现的，它们只能在同步块或同步方法中使用。wait() 方法使线程进入等待状态并释放锁。调用 sleep() 方法也可以使线程进入休眠状态，但并不释放锁。

The wait() method takes two forms:

(1) wait(time): pause for a specified period of time and enter into the ready state automatically.

(2) wait(): make a thread enter into the wait state, until notify() or notifyAll() is invoked to wake it up.

The prototypes of notify() and notifyAll() ara as below:

```
public final void notify()
public final void notifyAll()
```

wait() 方法有两种形式：

（1）wait(time): 暂停指定时间 time，时间到后自动进入就绪状态。

（2）wait(): 线程进入等待状态直到其他线程调用 notify() 或 notifyAll() 方法将其唤醒。

Only the thread that owns the lock of the object, that is, the method or a block of codes is modified with the keyword "synchronized" can call the notify() method to wake up a thread in the threads' waiting queue. The wait() and notify() are used in pairs, or else, a deadlock may phappen if a thread called wait() without the corresponding notify() to wake it up. To avoid this, it is usually recommended to use the notifyAll() method to wake up all threads in the waiting queue.

只有拥有对象锁的线程，即用 synchronized 修饰的方法或语句块才能调用 notify() 方法，并从 wait 等待队列中唤醒任一个线程。wait() 与 notify() 是配对使用的，若一个线程调用了 wait()，而未用相应的 notify() 去唤醒它，有可能会导致死锁。为了避免此种情况，通常建议使用 notifyAll() 方法唤醒等待队列中的所有线程。

Suppose two threads, P1 and P2, share the same data. P1 needs to access to it first and P2 next, the wait() and notify() methods can be used to satisfy the request.

```
synchronized void methodP1(…) { //thread P1 accesses to the shared data
  … //Codes access to the shared data
  avaliable=true;
  notify();
}
synchronized void methodP2(…) { //Thread P2 accesses to the shared data
  while(!avaliable) {
    try {
      wait();
    }
    catch(InterruptedException e) {
      …
    }
  }
  … //accesses to the shared data
}
```

假如两个线程 P1 和 P2 共享数据，要求 P1 先访问，P2 后访问，可由 wait() 和 notify() 方法实现。

Exercises

1. Write a program to print "How are you?" and "Fine, thank you!" three times respectively. Delay for a while after printing out one message.

(1) Define a subclass of Thread and override the run method of the Thread.

(2) In main class, create 2 objects (threads).

写程序，分别输出 "How are you?" 和 "Fine, thank you!" 3 次，每输出一个都延迟一点时间再输出。

（1）定义一个 Thread 的子类，重写 run 方法。

（2）在主类里创建 2 个对象（线程）并启动。

2. Write a program to create and start 6 threads.

(1) Define a subclass of Thread, named Test, which has an int k field and a constructor that takes an int parameter for initializing the k field.

(2) Inside run(), print a message "(I'm "+k+")", repeat this three times.

(3) Create and start 6 threads in main (create an array: for instance Test[] a=new Test[6]; …, a[i]=new Test(i);).

写程序，创建并启动 6 个线程。

（1）定义一个 Thread 的子类，命名为 Test，类里有个 int k，定义带一个 int 参数的构造方法，在构造方法里为 k 赋初值。

（2）在 run 方法里输出 "(I'm "+k+")" 3 次。

（3）在主类创建并启动这 6 个线程（创建对象数组，例：Test[] a=new Test[6]; …a[i]=new Test(i);）

3. Write a program, create threads indirectly. Define a class that implements Runnable interface. Inside run(), print a message(with 100 milliseconds delay), define a constructor and output a startup message to tell the sequence of the threads. Create 3 threads in main.

(1) Define a class, it implements Runnable interface.

(2) Define main class. In main create an object of the class defined in (1).

(3) In main, create three threads (objects) of the Thread class, and use the objects defined in (2) as the actual parameters of the Thread's constructor(public Thread(Runnable target);).

写程序，用间接法创建线程。定义一个实现 Runable 接口的类，在 run 方法里输出任意信息（延时 100 毫秒），定义构造方法，在其中输出线程的序号。在 main 里创建 3 个线程。

(1) 定义一个实现 Runable 接口的类。

(2) 定义主类，在 main 里创建 (1) 中定义的类的对象。

(3) 在 main 中创建 3 个线程（对象），用 (2) 中定义的对象作为 Thread 的构造方法的实参。

4. Redo exercise 1, with the indirect approach.

(1) Define a class to implement the interface Runnable.

(2) In main, create two Runnable objects of the class defined in (1) that will be the actual parameters of the Thread's constructor.

用间接法重写第 1 题。

(1) 定义一个实现 Runable 接口的类。

(2) 在 main 里创建 2 个在 (1) 中定义的类的对象，它们将作为 Thread 的构造方法的实参。

Chapter 11
Networking（网络）

11.1　Concept of networking（网络的概念）

In a network, different computers communicate each other and exchange information. In Java, basic networking is supported by classes in java.net package. These classes support connecting and retrieving imformation by HTTP and FTP protocols, also support lower level sockets communication. Using TCP or UDP, Java programs can communicate over the Internet.

在网络中，计算机之间可以交互信息。与网络通信相关的类在 Java.net 包里，利用这些类，可以实现 FTP 与 HTTP 协议的信息传输以及 sockets 通信。使用 TCP 或 UDP 协议，Java 程序可通过互联网在不同的计算机之间交换信息。

On the Internet, resources of data, audios and images are stored in files. A resource can be something as simple as a file or a directory. Java allows you to develop clients that retrieve files from a remote web server. The web server can be used to send files or other resources to the clients. Java supports developping server side programs to send files or resources to clients. The procedure of reading files from remote hosts to your Java program is as shown in Figure 11-1.

互联网上，数据、声音和图像等资源都以文件为单位存储。资源其实就像文件或者目录那样简单，可以用 Java 编写 clients 端程序，用以访问远程计算机上的文件（资源），由远程计算机的 Web 服务器将文件发送给客户。Java 也支持编写服务器端程序，向客户发送文件或资源。资源在网上的传输过程如图 11-1 所示。

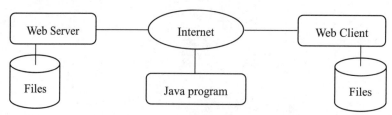

Figure 11-1 Transfer files on the Internet

11.2 URL class（URL 类）

The first thing to read files from the Internet is to locate the host computer where the files are. java.net.URL class can be used to identify the files on the Internet. In general, an URL (Uniform Resource Locator) is a pointer to a "resource" on the World Wide Web. As long as providing a URL, the files on the Internet can be located. Java programs that interact with the Internet also use URLs to find the resources on the Internet. A Java program uses class URL to provide an URL address.

In a Java program, an URL object represents an URL address. The URL class has several constructors. The prototype of a constructor is as such:

public URL(String spec) throws MalformedURLException

Where, spec represents an URL address. If the URL address is wrong, or can't be resolved, a MalformatedURLException exception will be thrown.

要从互联网读取文件，首要的是定位文件所在的计算机。java.net.URL 类可用于文件的定位。通常，把 URL (Uniform Resource Locator) 看作是 WWW 上的"资源"的地址，只要提供 URL，就可以定位文件。Java 程序与互联网的交互也要通过 URL 定位互联网上的资源来实现。

Java 程序利用 URL 类提供 URL 地址。Java 程序中，URL 类的对象代表 URL 地址，URL 类有很多构造方法。其中一个构造方法的原型如下：

```
public URL(String spec) throws MalformedURLException
```

其中，spec 是 URL 地址。如果 URL 地址不对，或者无法解析，会抛出一个 Malformated URLException 异常。

【Example 11-1】Create an URL object

```
import java.net.*;
class TryURL_1 {
  public static void main(String args[]) {
    try {
      URL url=new URL("http://www.sun.com");
    }
    catch(MalformedURLException ex) {}
  }
}
```

An URL has two main components:

(1) Protocol identifier.

In Example 11-1, the protocol identifier is http (Hypertext Transfer Protocol).

(2) Resource name.

The resource name is www.sun.com

The protocol identifier and the resource name are separated by a colon and two forward slashes.

Class URL has many useful methods, some of them are as shown in Figure 11-2.

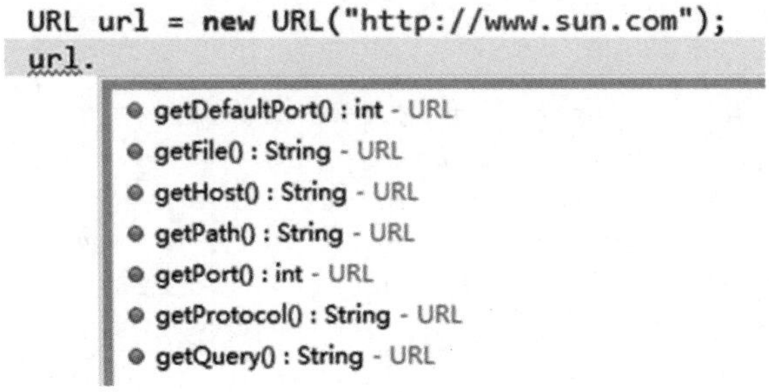

```
URL url = new URL("http://www.sun.com");
url.
        ● getDefaultPort() : int - URL
        ● getFile() : String - URL
        ● getHost() : String - URL
        ● getPath() : String - URL
        ● getPort() : int - URL
        ● getProtocol() : String - URL
        ● getQuery() : String - URL
```

Figure 11-2　A few methods of class URL

【Example 11-2】Create an URL object for http //www.sun.com

```
import java.net.*;
public class TryURL_2 {
  public static void main(String args[]) {
    try {
      URL url=new URL("http://www.sun.com");
      System.out.println(url.getProtocol());
      System.out.println(url.getHost());
      System.out.println(url.getRef());
      System.out.println(url.getDefaultPort());
    }
    catch(MalformedURLException ex) {}
  }
}
```

Output:

http

www.sun.com

null

80

If you want to read resources (files), just invoke openStream() method of the URL object. To retrieve the file of an Internet resource, you need to:

(1) Create an URL object for the file.

(2) Use the openStream() method of the URL class to open an input stream to the file's

URL.

```
URL url=new URL("http://www.cs.armstrong.edu/liang/index.html");
InputStream input=url.openStream();
```

Method openStream returns an object of the InputStream class.

【Example 11-3】**Read file from a web site**

```
import java.net.URL;
import java.io.*;
public class ReadSource {
  public static void main(String args[]) {
    int ch;
    try {
      URL url=new URL("http://mail.163.com");
      InputStream in=url.openStream();
      InputStreamReader rd=new InputStreamReader(in);
      while((ch=rd.read())!=-1)
        System.out.print((char)ch);
    }
    catch(IOException e) {
      System.out.println("Network error! ");
    }
  }
}
```

Output:

<html> //source code of the web page

<head><title>301 Moved Permanently</title></head>

<body bgcolor="white">

* <h1>301 Moved Permanently</h1> *

<hr><center>nginx</center>

</body>

</html>

【Example 11-4】**Use BufferedReader class**

```
import java.net.URL;
import java.io.*;
public class ReadSource {
  public static void main(String args[]) {
    String line;
    try {
      URL url=new URL("http://mail.163.com");
      InputStream in=url.openStream();                           //byte stream
      InputStreamReader rd=new InputStreamReader(in); //char stream
      BufferedReader bf=new BufferedReader(rd);
      while((line=bf.readLine())!=null) //public String readLine()
        System.out.println(line);
```

```
    }
    catch(IOException e) {
      System.out.println("Network error!");
    }
  }
}
```

11.3 Sockets communication（套接字通信）

URLs provide a relatively high-level mechanism for accessing resources on the Internet. Sometimes programs need lower-level network communication with sockets, for example, when you develop a client-server application. The client-server applications on the Internet can adopt TCP protocol. TCP provides a reliable, point-to-point communication.

A socket is the combination of an IP address and a port number. A socket is one endpoint of a two-way communication between two programs running on the network. The socket communication is based on the TCP protocol. A client program and a server program establish a connection to one another. Each program binds a socket to its connection, then the client and the server read from and write to the sockets bound to the connection. The objects of socket classes are used to represent the connection between a client program and a server program.

URLs 机制用于访问互联网上的资源。对于客户/服务器这样低层次的通信，需要用套接字实现。客户/服务器通信可以采用 TCP 协议，它提供可靠的点对点的有连接的通信。

套接字是 IP 地址和端口号的组合。套接字是在网络上运行的两个程序之间双向通信的端点。套接字通信可以采用 TCP 协议。为了实现客户程序与服务器程序的通信，客户端和服务器端都要设计程序，通信双方通过套接字建立连接及读/写数据。Socket 类用于客户/服务器程序之间建立连接与交换数据。套接字（socket）为两台计算机之间的通信提供了一种机制，客户和服务器分别读取和写入数据到套接字，套接字实现客户与服务器的连接。

11.3.1 How Socket communication works?（套接字通信是如何进行的？）

Normally, a server running on a computer has a socket that is bound to a specific port number to tell the system to receive all data through that port. The client also needs to bind to a local port number that will be used during connecting with the server.

In the socket communication, the server program keeps running and listening to the socket for a client to make a connection request. The clients request for connections to the server. After the server accepts the request, the communication between the server and the client is conducted (read/write data). In nfact, the information transmition through internet is an I/O operation.

把服务器计算机的 socket 绑定到一个特定的端口，用于通知系统，将从这个端口接收数据。客户计算机也要绑定到一个端口，用于与服务器的连接。

在套接字（socket）通信中，服务器程序要先运行，等待客户机的连接请求。客户机发出连接请求，服务器接受请求后，服务器与客户机便开始通信（读/写数据），通信过程即是 I/O 过程。

11.3.2　Ports（端口）

A socket needs to bind to a port. Usually, a computer has a single physical connection to the network. All data arrives through the connection. By ports, the computer knows to which application the data forwards.

The TCP and UDP protocols use ports to map incoming data to a particular process running on a computer. Data transmits over the Internet using the address information URL to locate a computer, and then using the port to determine the data to and from which program (application).

In fact, the computer is identified by its 32-bit IP address, and ports are identified by a 16-bit number, with which TCP delivers data to the right application.

The 16-bit means the numbers ranging from 0 to 65535. The port numbers ranging from 0-1023 are reserved for services such as HTTP and FTP and other system devices.

Socket 要绑定到一个端口。通常，计算机与网络只有一个物理连接，网上的数据都是通过这个物理连接到达计算机的。通过端口，网上的数据能够送到指定的应用（程序）。

TCP 和 UDP 协议利用端口将数据发送给计算机上正在运行的特定的应用。URL 用于定位计算机，实现数据在互联网上的传输，端口则定位具体的应用程序。

事实上，识别计算机是通过一个 32 位的 IP 地址实现的。识别应用（程序）则是通过一个 16 位的整数（端口号）实现的。

16 位能够表示的数的范围为 0～65535。端口号 0～1023 为系统保留，例如 HTTP 和 FTP 以及其他系统设备的端口号。

11.4　ServerSocket and Socket classes（ServerSocket 和 Socket 类）

In Java.net package there are Socket and ServerSocket classes. The Socket class implements the client side socket of a two way connection between a Java program and other program on the network. The ServerSocket class implements the socket that servers can use to listen and accept connections to clients. The prototypes of the constructors are as follows:

```
public ServerSocket(int port)
public Socket(InetAddress address, int port)
```

在 java.net 包里有 Socket 和 ServerSocket 类。Socket 类封装了通信双方的客户端的操作，

ServerSocket 类封装了服务器端的操作。

For instance:

```
ServerSocket Listen=new ServerSocket(4321);        //Server side
Socket service=new Socket("Email server",4321);    //Client side
```

Figures 11-3 and 11-4 show some of methods the two classes.

```
ServerSocket server=new ServerSocket(4321);
server.
```
- accept() : Socket - ServerSocket
- bind(SocketAddress arg0) : void - ServerSocket
- bind(SocketAddress arg0, int arg1) : void - ServerS
- close() : void - ServerSocket

Figure 11-3 Some methods of class ServerSocket

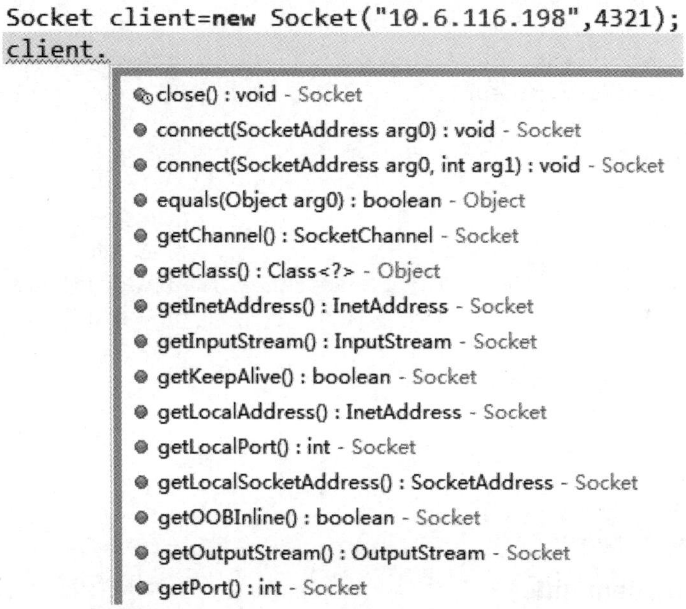

```
Socket client=new Socket("10.6.116.198",4321);
client.
```
- close() : void - Socket
- connect(SocketAddress arg0) : void - Socket
- connect(SocketAddress arg0, int arg1) : void - Socket
- equals(Object arg0) : boolean - Object
- getChannel() : SocketChannel - Socket
- getClass() : Class<?> - Object
- getInetAddress() : InetAddress - Socket
- getInputStream() : InputStream - Socket
- getKeepAlive() : boolean - Socket
- getLocalAddress() : InetAddress - Socket
- getLocalPort() : int - Socket
- getLocalSocketAddress() : SocketAddress - Socket
- getOOBInline() : boolean - Socket
- getOutputStream() : OutputStream - Socket
- getPort() : int - Socket

Figure 11-4 Some methods of class Socket

11.4.1 Tasks of each side（双方的任务）

Figure 11-5 shows the tasks of the server program and the client program in the socket communication.

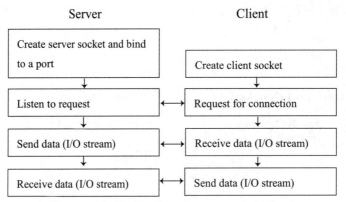

Figure 11-5 Tasks of server and client

1. Socket in server side

```
…
try {
  ServerSocket srv=new ServerSocket(2000);   //create server socket
  Socket socket=srv.accept();                //listen to request
}
catch (IOException e) {}
…
                                             //Read text from the socket
try {                                        //receive data
  BufferedReader rd=new BufferedReader(new InputStreamReader(socket.getInputStream()));
  String str;
  while ((str=rd.readLine())!=null) {
    System.out.println(str);
  }
  rd.close();
}
catch (IOException e) {}
```

2. Socket in client side

```
…
try {
  Socket socket=new Socket(IPaddr, 2000);   //create socket and request for connection
}
catch (Exception e) {}
…
try {
  OutputStreamWriter wr=new OutputStreamWriter(socket.getOutputStream());
  wr.write("Hello\r\n");                    //send data
}
catch (IOException e) {}
```

11.4.2 Data transmission（数据传输）

```
ServerSockets svr=new ServerSocket(8000);
```

```
Socket socket=svr.accept(); Socket socket=new Socket("ServerIP", 8000);
```

The server program and the client program make the connection through two sockets are as shown in Figure 11-6.

Receive data:
InputStream input=socket.getInputStream()
Send data:
OutputStream opt=socket.getOutputStream()

Send data:
OutputStream opt=socket.getOutputStream()
Receive data:
InputStream input=socket.getInputStream()

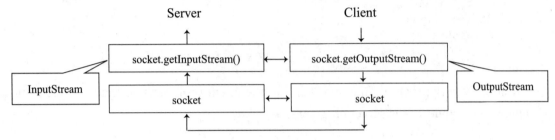

Figure11-6　Communication through sockets

【Example 11-5】**Server receives data from clients**

```
import java.io.*;
import java.net.*;
class Server {
  public static void main(String[] arg) {
    int count=0; //count number of clients
    byte b[]=new byte[100];
    InputStream in;
    int k;
    try {
      ServerSocket server=new ServerSocket(9876);
      while(true) { //wait for many clients
        Socket sc=server.accept();
        in=sc.getInputStream();
        k=in.read(b);
        System.out.print("Client message:");
        for(int i=0;i<k;i++)
          System.out.print((char)b[i]);
        System.out.println();
        System.out.println("This is "+(++count)+" client");
        System.out.println("Client IP: "+sc.getInetAddress());
        System.out.println("port: "+sc.getLocalPort());
        in.close();
        sc.close();
      }
    }
    catch(Exception e){}
  }
}
```

【Example 11-6】Client sends data to server

```java
import java.io.*;
import java.net.*;
public class Client {
  public static void main(String[] args) throws Exception {
    byte[] a={'H','e','l','l','o'};
    OutputStream dout;
    Socket sc=new Socket("localhost",9876);
    dout=sc.getOutputStream();
    dout.write(a);
    dout.close();
    sc.close();
  }
}
```

Run the server program first and then run the client program two times on the same computer, the outputs of the server program is as below.

 Output:

Client message:Hello
This is 1 client
Client IP:/127.0.0.1
port:9876
Client message:Hello
This is 2 client
Client IP:/127.0.0.1
port:9876

It's inconvenient to send messages in bytes one by one, so you can define a byte array to access data in groups. The following two examples send and receive data with byte arrays.

【Example 11-7】Server receives and sends messages to and from the client. Read first!

```java
import java.io.*;
import java.net.*;
public class TcpServer {
  public static void main(String[] args) throws Exception {
    byte[] buffer=new byte[200];
    InputStream is;
    OutputStream os;
    int length=0;
    ServerSocket ss=new ServerSocket(5000);
    while(true) {
      Socket socket=ss.accept();
      is=socket.getInputStream();
      os=socket.getOutputStream();
      length=is.read(buffer);                    //data read stored in buffer
      String str=new String(buffer,0,length);    //convert byte to string
```

```
            System.out.println(str);
            os.write("Client, How are you?".getBytes()); //convert string to byte
            is.close();
            os.close();
            socket.close();
        }
    }
}
```

【Example 11-8】 Client sends and receives messages to and from the server. Write first!

```
import java.io.*;
import java.net.*;
public class TcpClient {
    public static void main(String[] args) throws Exception {
        Socket socket=new Socket("localhost",5000);
        OutputStream os=socket.getOutputStream();
        os.write("hello world".getBytes());
        InputStream is=socket.getInputStream();
        byte[] buffer=new byte[200];
        int length=is.read(buffer);
        String str=new String(buffer,0,length);
        System.out.println(str);
        is.close();
        os.close();
        socket.close();
    }
}
```

A socket can only access to byte streams, however, most commonly, we like to access to char streams, even strings. Some I/O classes introduced in chapter nine can be used for the conversion. By an InputStreamReader object, the byte stream from a socket can be converted to a char stream. By an OutputStreamWriter object, a char stream can be converted to a byte stream.

For instance:

```
InputStreamReader in=new InputStreamReader(socket.getInputStream());
int k=in.read(b); //k:number of characters read, b:a char array
OutputStreamWriter out=new OutputStreamWriter(socket.getOutputStream());
out.write("I'm a client\r\n");
```

You may like to send any messages from the keyboard instead of a fixed one. The System.in is the object corresponding to the keyboard, also by an InputStreamReader object, the byte stream from the keyboard can be converted to the char stream, and further, by a BuffedReader object, a method readLine() can be used to read a line from the keyboard.

```
BufferedReader stdin=new BufferedReader(new InputStreamReader(System.in));
String str=stdin.readLine();
```

Socket 只能发送 / 接收字节流，然而，我们常常需要发送 / 接收字符流，甚至字符串。第 9 章的 I/O 类可以用于字节流与字符流的转换。InputStreamReader 类可以将来自 socket 的字节流转换成字符流，也可以利用 OutputStreamWriter 类将字符流转换成字节流，再通过 socket 发送到网上。如果需要发送来自键盘的任何数据，而不是固定的信息，也可以利用 InputStreamReader 类进行转换。System.in 对象代表键盘输入，通过 InputStreamReader 对象将其转换成字符流，再通过 BuffedReader 对象的 readLine 方法，从键盘读取一行字符。

【Example 11-9】**Server receives messages**

```
import java.io.*;
import java.net.*;
class Server {
  public static void main(String[] arg) {
    int count=0;       //count number of clients
    char b[]=new char[100];
    InputStreamReader in;
    int k;
    try {
      ServerSocket server=new ServerSocket(9876);
      while(true) {
        Socket sc=server.accept();
        in=new
        InputStreamReader(sc.getInputStream());
        //convert byte stream to char stream
        k=in.read(b);  //number of characters
        System.out.print("Client message:");
        for(int i=0;i<k;i++)
        System.out.print(b[i]);
        System.out.println();
        System.out.println("This is "+(++count)+" client");
        System.out.println("Client IP:"+sc.getInetAddress());
        System.out.println("port:"+sc.getLocalPort());
        in.close();
        sc.close();
      }
    }
    catch(Exception e){}
  }
}
```

After received a message from a client program, the following results come up.

 Output:

I'm a client
This is 1 client
Client IP:/127.0.0.1
port:9876

【Example 11-10】**Client sends messages to the server**

```
import java.io.*;
import java.net.*;
```

```java
public class Client {
  public static void main(String[] args) throws Exception {
    Socket sc=new Socket("localhost",9876);
    OutputStreamWriter out=new OutputStreamWriter(sc.getOutputStream());
    out.write("I'm a client\r\n");
    out.close();
    sc.close();
  }
}
```

11.5 Serving multiple clients（服务多个客户）

Multiple clients are quite often to connect to a single server at the same time. Usually, a server runs constantly on a server computer, and clients from all over the Internet may want to connect to it. You can use threads to handle the server's multiple clients simultaneously. The server creates two threads (read and write) for each connection. Here is how the server handles a connection.

通常多个客户会同时访问一台服务器。互联网上的客户都想要访问服务器，所以服务器程序会一直运行。通过线程，服务器可以同时处理多个客户的访问。服务器会为每个与之连接的客户创建2个线程（读和写）。下面代码演示服务器如何处理一个连接。

1. Server side

【Example 11-11】**Simple talk programs**

```java
import java.net.*;
import java.io.*;
public class TalkServer {
  public static void main(String arg[]) {
    try {
      ServerSocket s=new ServerSocket(5432);
      while(true) {              //wait for many clients
        Socket s1=s.accept();    //wait for client connection
        new Soutput(s1).start(); //for each client create two threads
        new Sinput(s1).start();
        System.out.println("Talk to clients:");
      }
    }
    catch(Exception e){System.out.println("server Error");}
  }
}
```

Thread for read:

```java
class Sinput extends Thread {
  private Socket socket;
  public Sinput(Socket sock) {socket=sock;}
  public void run() {
    String str="";
    try {
```

```
        BufferedReader is; //read from client
        is=new BufferedReader(new InputStreamReader(socket.getInputStream()));
        while(true) {
          str=is.readLine();
          System.out.print("This is client"+socket.getInetAddress()+" He
          or She says ");
          System.out.println(str);
        }
      }
      catch(IOException e) {e.printStackTrace();}
  }
}
```

Thread for keyboard input and write:

```
class Soutput extends Thread {
  private Socket socket;
  public Soutput(Socket sock){socket=sock;}
  public void run() {
    try {
      OutputStreamWriter os; //write to server
      os=new OutputStreamWriter(socket.getOutputStream());
      BufferedReader reader; //read from keyboard
      reader=new BufferedReader(new InputStreamReader(System.in));
      while(true) {
        String line=reader.readLine();
        line=line+"\r\n";
        os.write(line);
        os.flush();
      }
    }
    catch(IOException e) { e.printStackTrace();}
  }
}
```

2. Client side

```
import java.net.*;
import java.io.*;
public class TalkClient {
  public static void main(String arg[]) throws IOException {
    Socket s1=new Socket("localhost",5432); //replace "localhost" with the server IP
    new Cinput(s1).start();
    new Coutput(s1).start();
    System.out.println("Talk to server:");
  }
}
```

Thread for read:

```
class Cinput extends Thread {
  private Socket socket;
  public Cinput(Socket sock) {
```

```
    super(); //can omit
    socket=sock;
  }
  public void run() {
    try {
      BufferedReader is;
      is=new BufferedReader(new InputStreamReader(socket.getInputStream()));
      while(true) {
        String str=is.readLine();
        System.out.print("The Server says ");
        System.out.println(str);
      }
    }
    catch(IOException e){e.printStackTrace();}
  }
}
```

Thread for keyboard input and write:

```
class Coutput extends Thread {
  private Socket socket;
  public Coutput(Socket sock) {socket=sock;}
  public void run() {
    try {
      OutputStreamWriter os;
      os=new OutputStreamWriter(socket.getOutputStream());
      BufferedReader reader;
      reader=new BufferedReader(new InputStreamReader(System.in));
      while(true) {
        String line=reader.readLine();
        line=line+"\r\n";
        os.write(line);
        os.flush();
      }
    }
    catch(IOException e){e.printStackTrace();}
  }
}
```

Exercises

1. Write two programs, a server side program and a client side one to send/receive messages. The client receives one message from the server and the server sends one message to the client.

2. Write two programs, a server side program and a client side one to send/receive messages. The client sends one message to the server and the server receives one message from the client.

3. Run the following two programs on two computers.

Hint:

(1) Acquire the server IP by command "ipconfig", and tell it to the other client computer.

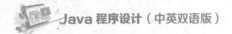

(2) Replace "localhost" in program Ct.java with the server's IP.

```java
import java.io.*;
import java.net.*;
class Svr {
  public static void main(String[] arg) {
    int count=0; //count number of clients
    try {
      ServerSocket server=new ServerSocket(9876);
      while(true) {
        Socket sc=server.accept();
        OutputStreamWriter out=new OutputStreamWriter(sc.getOutputStream());
        System.out.println("This is "+(++count)+" client");
        System.out.println("IP: "+sc.getInetAddress());
        System.out.println("port: "+sc.getLocalPort());
        out.write("This is Server.\r\n");
        out.write("Who are you?\r\n");
        out.write("Bye!\r\n");
        out.close();
        sc.close();
      }
    }
    catch(Exception e){System.out.println("Error");}
  }
}
```

```java
import java.io.*;
import java.net.*;
public class Ct {
  public static void main(String[] args) throws Exception {
    String str="";
    Socket sc=new Socket("localhost",9876); //replace "localhost" with the server IP
    BufferedReader din=new BufferedReader(new InputStreamReader(sc.getInputStream()));
    while(true) {
      str=din.readLine();
      if(str!=null)
        System.out.println("Client:"+str);
      else
        break;
    }
    din.close();
    sc.close();
  }
}
```

4. Write programs for a simple talk. The server side program is as below, write the client side program.

```java
import java.net.*;
```

```java
import java.io.*;
public class TalkServer {
  public static void main(String arg[]) {
    try {
      ServerSocket s=new ServerSocket(5432);
      while(true) {                  //wait for many clients
        Socket s1=s.accept();        //wait for client connection
        new Soutput(s1).start();     //for each client create 2 threads
        new Sinput(s1).start();
      }
    }
    catch(Exception e){System.out.println("server Error");}
  }
}
class Sinput extends Thread {
  private Socket socket;
  public Sinput(Socket sock) {socket=sock;}
  public void run() {
    String str="";
    try {
      BufferedReader is;
      is=new BufferedReader(new InputStreamReader(socket.getInputStream()));
      while(true) {
        str=is.readLine();
        System.out.println(str);
      }
    }
    catch(IOException e) {e.printStackTrace();}
  }
}
class Soutput extends Thread {
  private Socket socket;
  public Soutput(Socket sock) {socket=sock;}
  public void run() {
    try {
      OutputStreamWriter os;
      os=new OutputStreamWriter(socket.getOutputStream());
      BufferedReader reader;
      reader=new BufferedReader(new InputStreamReader(System.in));
      while(true) {
        String line=reader.readLine();
        line=line+"\r\n";
        os.write(line);
        os.flush();
      }
    }
    catch(IOException e) {e.printStackTrace();}
  }
}
```

References

[1] 张勇. Java 程序设计与实践教程 [M]. 北京：人民邮电出版社，2014.

[2] ECKEL. Java 编程思想 [M]. 陈吴鹏，译. 北京：机械工业出版社，2015.

[3] 明日科技. Java 从入门到精通 [M]. 北京：清华大学出版社，2012.

[4] 叶核亚. Java 程序设计实用教程 [M]. 5 版. 北京：电子工业出版社，2019.

[5] 谷志峰，琚伟伟. Java 程序设计基础教程 [M]. 北京：电子工业出版社，2016.